SpringerBriefs in Mathematics

SpringerBriefs in Mathematics showcases expositions in all areas of mathematics and applied mathematics. Manuscripts presenting new results or a single new result in a classical field, new field, or an emerging topic, applications, or bridges between new results and already published works, are encouraged. The series is intended for mathematicians and applied mathematicians.

More information about this series at http://www.springer.com/series/10030

Gary Chartrand · Teresa W. Haynes ·
Michael A. Henning · Ping Zhang

From Domination to Coloring

Stephen Hedetniemi's Graph Theory and Beyond

 Springer

Gary Chartrand
Department of Mathematics
Western Michigan University
Kalamazoo, MI, USA

Michael A. Henning
Department of Mathematics
University of Johannesburg
Johannesburg, South Africa

Teresa W. Haynes
Department of Mathematics
East Tennessee State University
Johnson City, TN, USA

Ping Zhang
Department of Mathematics
Western Michigan University
Kalamazoo, MI, USA

ISSN 2191-8198   ISSN 2191-8201 (electronic)
SpringerBriefs in Mathematics
ISBN 978-3-030-31109-4   ISBN 978-3-030-31110-0 (eBook)
https://doi.org/10.1007/978-3-030-31110-0

This Springer imprint is published by the registered company Springer Nature Switzerland AG
The registered company address is: Gewerbestrasse 11, 6330 Cham, Switzerland

Dedicated to

Stephen T. Hedetniemi
On the Occasion of his 80th Birthday
February 7, 2019

Preface

As with every area of mathematics, graph theory has a number of mathematicians who have contributed to its development in a number of ways, namely (1) by proving theorems that are instrumental in its growth, (2) by giving lectures and writing survey papers and books that popularize graph theory, and (3) by creating new concepts and topics that have drawn mathematicians into various areas of graph theory. One mathematician responsible for all of this is Stephen T. Hedetniemi. Steve earned his Ph.D. in mathematics with a specialization in graph theory at the University of Michigan in 1966 under the direction of the well-known graph theorist Frank Harary.

Two major areas of research by Steve Hedetniemi are domination and coloring. In this book, we begin by discussing several topics, results, and problems in domination in which Steve has made a major contribution. From domination, we move on to a number of coloring topics. Along the way *from domination to coloring*, we also discuss other research topics in Stephen Hedetniemi's graph theory, including distance in graphs and two types of traversing walks. In the eight chapters that follow, while the material presented represents only a small sample of Steve's research in graph theory, we believe that beyond what is included lies other avenues for research.

Through studying chessboard problems, Stephen Hedetniemi introduced total domination, which has become one of the major topics of study in domination. Hedetniemi and others showed that there is a chain of inequalities involving the domination number of a graph, the independent domination number, and other domination-related parameters. These are the primary topics of Chap. 1. The independent domination number and total domination number are discussed in more detail in Chap. 2. If every vertex in a dominating set S of a graph G is assigned the value 1 and the vertices not in S are assigned 0, then the sum of the values of each vertex of G and its neighbors is at least 1. This observation by Hedetniemi led to the introduction of a dominating function of a graph. This concept, together with some variations, is the subject of Chap. 3. Two recent domination-related parameters introduced by Hedetniemi, namely Roman domination and alliances in graphs, are the subject of Chap. 4. In the first four chapters then, we discuss some of the primary and most recent results dealing with prominent domination parameters, as well as new and interesting concepts and problems derived from these concepts.

Many areas of graph theory different from domination have also been influenced by the research of Stephen Hedetniemi. One of these is distance in graphs in which Steve has investigated two interpretations of the "middle" of a graph, namely the center and median, which have numerous applications. These and other distance-related subgraphs of graphs are the topics of Chap. 5. In Chap. 6, we discuss two graph traversing concepts studied by Hedetniemi and his coauthors, one in which all edges of a graph are traversed, resulting in Eulerian walks, and a second in which all vertices are traversed, resulting in Hamiltonian walks.

Graph colorings has been a popular area of research for well over a century. This has also been a topic of interest for Hedetniemi for many years. In fact, he wrote his doctoral dissertation on graph homomorphisms, a concept closely tied to proper colorings of graphs. The concept of graph homomorphisms occurs in both Chaps. 7 and 8. Every proper coloring of a graph using the minimum number of colors has the property that for every two distinct colors, there are adjacent vertices with these colors. Any coloring with this property is a complete coloring, which is the primary topic of Chap. 7. The two major methods of evaluating how highly connected a graph involves vertex-cuts and edge-cuts. In Chap. 8, we see relationships of these concepts with graph colorings, resulting in color connection and disconnection in graphs. Recent results involving these connectivity-coloring concepts are presented along with suggestions for new avenues of research.

Kalamazoo, MI, USA   Gary Chartrand
Johnson City, TN, USA   Teresa W. Haynes
Johannesburg, South Africa   Michael A. Henning
Kalamazoo, MI, USA   Ping Zhang
August 2019

Contents

Chapter 1
Pioneer of Domination in Graphs

Stephen Hedetniemi is perhaps best known for his pioneering work in domination in graphs. In this chapter, we explore some of his contributions to the direction and advancement of this field of study. We focus on two topics, namely domination of chessboard graphs and the domination chain.

1.1 Introduction

Honest pioneer work in the field of science has always been, and will continue to be, life's pilot. Wilhelm Reich

Pioneering is the work of individuals. Susanne Katherina Langer

Stephen Hedetniemi is at the top of the list of individuals who have most influenced the growth of the popular area of domination in graphs. In this chapter, we first discuss the origin of domination as a chessboard covering problem and consider Steve's contribution to this area of study. Then we turn our attention to the "so-called" domination chain, which was introduced by Hedetniemi along with Cockayne and Miller.

In the subsequent two sections, we will use the following terminology and introduce additional notation as needed. A set S of vertices of a graph G is *independent* if no two vertices in S are adjacent, and the maximum cardinality of an independent set of G is the *independence number* of G, denoted $\alpha(G)$. A *dominating set* S of G is a set of vertices of G such that every vertex in $V \setminus S$ is adjacent to a vertex in S, and the *domination number* $\gamma(G)$ is the minimum cardinality of a dominating set of G. The *independent domination number* of G, denoted $i(G)$, is the minimum cardinality of an independent dominating set of G.

© The Author(s), under exclusive license to Springer Nature Switzerland AG 2019
G. Chartrand et al., *From Domination to Coloring*, SpringerBriefs in Mathematics,
https://doi.org/10.1007/978-3-030-31110-0_1

1.2   Covering Chessboards

Play is the highest form of research. Albert Einstein

Although domination in graphs was not formally defined in mathematics until the 1960s, its inception occurred in the 1860s under the guise of covering a chessboard. A chess piece is said to *cover* (attack) any square on a chessboard that it can reach in a single move. For example, in one move a queen can move any number of squares horizontally, vertically, or diagonally. Thus, a queen covers the squares in the same row, column, or diagonal with it.

In 1862, de Jaenisch [13] posed the problem of determining the minimum number of queens required to cover an $n \times n$ chessboard. To interpret a question of this type as a graph theory problem, a graph is formed by representing each square as a vertex, with two vertices adjacent if a chess piece positioned on one square covers the other. In particular, the Queen's graph, denoted Q_n, for an $n \times n$ chessboard has n^2 vertices, where two vertices are adjacent if and only if the squares they represent are in the same row, column, or diagonal. It follows that de Jaenisch's question, posed some one hundred years prior to the formalization of domination in graphs, translates to determining the domination number of the Queen's graph.

From the perspective of chessboard coverings, de Jaenisch [13] determined minimum dominating sets and minimum independent dominating sets of Q_n for $n \leq 8$. In particular, he asserts that $\gamma(Q_8) = 5$. See Fig. 1.1 for an example of a minimum cover of the 8×8 chessboard with five queens. For another example, the placement of queens shown in Fig. 1.2 covers the board with the added constraint that no two

Fig. 1.1 Five queens covering an 8×8 chessboard

Fig. 1.2 Minimum
independent dominating
set of queens

queens can attack (cover) each other. That is, these five queens represent a minimum
independent dominating set of the Queen's graph and $\gamma(Q_8) = i(Q_8) = 5$.

W. W. Rouse Ball [2] listed three basic types of problems being studied on chessboards in 1892 as follows:

1. Covering: Determine the minimum number of chess pieces of a given type that
   are required to cover every square of an $n \times n$ chessboard. (Note that this is de
   Jaenish's question when the specified chess piece is a queen.)
2. Independent Covering: Determine the minimum number of mutually nonattacking
   chess pieces of a given type that are required to cover every square of an $n \times n$
   chessboard.
3. Independence: Determine the maximum number of chess pieces of a given type
   that can be placed on an $n \times n$ chessboard such that no two pieces attack (cover)
   each other.

Hence, for the graph G associated with the given chess piece and chessboard, the
problems translate to determining (1) the domination number $\gamma(G)$, (2) the independent domination number $i(G)$, and (3) the independence number $\alpha(G)$. If the
chess piece being considered is the queen, Problem (3) is commonly known as the
N-queens Problem. A solution to the N-queens Problem showing that $\alpha(Q_8) = 8$
is shown in Fig. 1.3. People have tried for over 150 years to find answers to these
types of problems on chessboards, unaware for the first hundred years or so that
they were actually trying to determine the domination, independent domination, and
independence numbers of chessboard graphs.

It is not surprising that as a pioneer of domination in graphs, Hedetniemi is also
interested in chessboard problems. In fact, Hedetniemi states in [18] that it was the

Fig. 1.3 Maximum
independent set of queens

solution to covering an 8×8 chessboard with five queens shown in Fig. 1.1 that led
to his discovering the total domination number. He noticed that in this particular
covering each of the five queens was covered by another queen. In other words, for
the associated dominating set S of the Queen's graph, not only did the vertices in
$V \setminus S$ have a neighbor in S, each vertex in S also had a neighbor in S. The realization
that this was a new type of domination that had not yet been studied led to the
introduction of total domination by Cockayne, Hedetniemi, and Dawes in their now
classic paper [8]. This paper spurred much growth in the field of domination as a
whole.

Also, noting that the queens in Fig. 1.1 are placed along a diagonal, Cockayne
and Hedetniemi [11] proposed a new variant of covering all the squares of an $n \times n$
chessboard by placing queens only on the main diagonal. Another interesting property
of this particular dominating set of queens is that all the queens are on squares of
the same color (with the usual alternating color scheme for board squares). In fact,
many minimum dominating sets of the Queen's graph are monochromatic. For more
examples of monochromatic dominating sets, a minimum independent dominating
set of the 11×11 chessboard is shown in Fig. 1.4 and a minimum dominating set
for the 13×13 chessboard is shown in Fig. 1.5.

Hedetniemi's research in chessboard problems extends to graphs defined by
the movements of various chess pieces in addition to the queen, for examples the
bishop, rook, and knight. In addition, Steve's promotion of chessboard type problems includes two excellent surveys, namely a 1995 survey [15] by Fricke, Hedetniemi, Hedetniemi, McRae, Wallis, Jacobson, Martin, and Weakley and a 1998 survey

Fig. 1.4 Five queens covering an 11 × 11 chessboard

chapter [17] by Hedetniemi, Hedetniemi, and Reynolds. These surveys stimulated interest and escalated research in the area of domination of chessboard graphs.

Douglas Weakley, a leading researcher in domination of the Queen's graph, credits Hedetniemi for his interest in chessboard problems. For a sample of Weakley's work on dominating the Queen's graph, see [14, 22–24]. In his chapter titled "Queens Around the World in Twenty-five Years", Weakley [25] states that Hedetniemi's fascinating lecture in the summer of 1991 at Indiana University–Purdue University Fort Wayne on chessboard graphs and domination parameters inspired him to change his research focus to this topic. So it seems that Steve has positively influenced Weakley, like he has so many others.

Determining the domination number of the Queen's graph appears to be a complex challenge. Anne Sinko and Peter Slater [21] describe it as a "long studied, highly entertaining, and very difficult problem". Today much of the research involved in determining exact values for chessboard type parameters is algorithmic in nature. As noted in [18, 20], only relatively few exact values of the domination number of the Queen's graph are known. According to Hedetniemi [18], the 2001 paper by Östergård and Weakley [20] is the definitive paper on the subject. The value of $\gamma(Q_n)$ is either known, or known to be one of two consecutive values for all $n \leq 120$ (see [20]). The known values of $\gamma(Q_n)$ and $i(Q_n)$ for $4 \leq n \leq 24$ are summarized

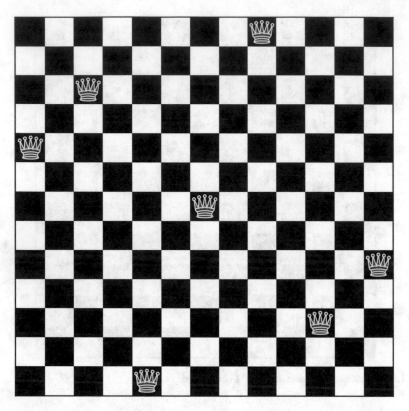

Fig. 1.5 Seven queens covering a 13 × 13 chessboard

Table 1.1 Values of $\gamma(Q_n)$ and $i(Q_n)$ for small n

n	4	5	6	7	8	9	10	11	12	13	14
$\gamma(Q_n)$	2	3	3	4	5	5	5	5	6	7	8
$i(Q_n)$	3	3	4	4	5	5	5	5	7	7	8
n	15	16	17	18	19	20	21	22	23	24	
$\gamma(Q_n)$	9	9	9	9	10	11	11	12	12	13	
$i(Q_n)$	9	9	9	10	11	11	11	12	13	13	

in Table 1.1 (see [4, 20]). The table values for $i(Q_n)$ for $n \in \{19, 20, 22, 23, 24\}$ and $\gamma(Q_n)$ for $n \in \{20, 22, 24\}$ are due to the 2017 doctoral dissertation of William Bird [4].

Covering with queens has also been studied for rectangular shaped boards, see [6] for example. Let $Q_{m \times n}$ denote the Queen's graph on the $m \times n$ chessboard. Figure 1.6 illustrates the unique (up to symmetry) minimum independent dominating set of $Q_{12 \times 18}$.

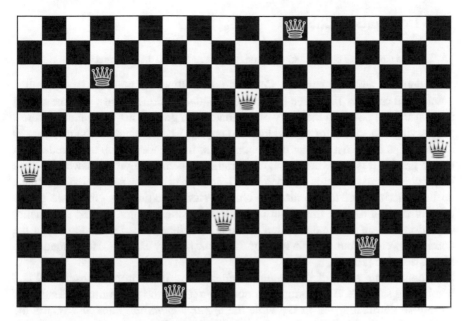

Fig. 1.6 Unique minimum independent dominating set of a 12×18 chessboard

We conclude this section with a conjecture made by Hedetniemi at the 1992 Western Michigan University Graph Theory Conference.

Conjecture 1.2.1 (Hedetniemi) *For any* $n \geq 1$, $\gamma(Q_n) \leq \gamma(Q_{n+1})$.

This conjecture remains open and is listed among Steve's top 10 favorite conjectures in [18]. Recent work of Bozóki, Gál, Marosi, and Weakley [6] shows that Conjecture 1.2.1 does not extend to rectangular boards. Note that $\gamma(Q_{m \times n})$ is the function of two variables, m and n. Hedetniemi's conjecture can be rephrased for each variable as follows: (1) Is $\gamma(Q_{m \times n})$ monotonic in m? That is, does $\gamma(Q_{m \times n}) \leq \gamma(Q_{m \times (n+1)})$ hold for $m \leq n$? and (2) Is $\gamma(Q_{m \times n})$ monotonic in n? That is, does $\gamma(Q_{m \times n}) \leq \gamma(Q_{(m+1) \times n})$ hold for $m \leq n$? The answer, as shown in [6], is negative for $\gamma(Q_{m \times n})$ in the first variable. In the range $1 \leq m \leq n \leq 18$, monotonicity in m of $\gamma(Q_{m \times n})$ fails once: $\gamma(Q_{8 \times 11}) = 6 > 5 = \gamma(Q_{9 \times 11}) = \gamma(Q_{10 \times 11}) = \gamma(Q_{11 \times 11})$. In the same range, monotonicity in m of $i(Q_{m \times n})$ fails twice: $i(Q_{8 \times 11}) = 6 > 5 = i(Q_{9 \times 11}) = i(Q_{10 \times 11}) = i(Q_{11 \times 11})$ and $i(Q_{11 \times 18}) = 9 > 8 = i(Q_{12 \times 18})$. It is noted in [6] that the failure of monotonicity seems to be due to a "special" minimum dominating set of the large board that does not fit on the small board. For $Q_{8 \times 11}$ versus $Q_{9 \times 11}$ through $Q_{11 \times 11}$, that special dominating set is shown in Fig. 1.4; and for $Q_{11 \times 17}$ versus $Q_{12 \times 18}$, the special set is shown in Fig. 1.6. Insight from this study of rectangular boards suggests that possibly what Conjecture 1.2.1 is really asking is whether there is an n for which Q_{n+1} has a special minimum dominating set that does not fit on Q_n.

1.3 Domination Chain

"Like the pioneers of old, a creative person breaks new ground daily." Anna Olson

Hedetniemi and his coauthors, Cockayne and Miller, first presented the so-called domination chain in 1978 (see [12]). This inequality sequence has become one of the major focal points in the study of domination in graphs, inspiring much interest and serving as a source for several hundred papers. The domination chain expresses relationships that exist among dominating sets, independent sets, and irredundant sets in graphs. It is noteworthy that the parameters being sought to answer the three major chessboard problems stated in Sect. 1.2 are precisely the parameters that make up the inner core of the domination chain. Prior to stating the chain, we give a brief discussion of how this chain is developed using maximality and minimality conditions. Much of this discussion is taken from the detailed development of the domination chain found in Chap. 3 of the book [16] by Haynes, Hedetniemi, and Slater. See also [10].

A dominating set S of G is *minimal dominating* if no proper subset of S is a dominating set of G. The domination number $\gamma(G)$ is the minimum cardinality of a minimal dominating set in G; while the *upper domination number* $\Gamma(G)$ is the maximum cardinality of a minimal dominating set. An independent set S of G is *maximal independent* if no proper superset of S is an independent set. Recall that $i(G)$ is the independent domination number and $\alpha(G)$ is the independence number of G. Thus, $i(G)$ is the minimum cardinality of a maximal independent set of G and $\alpha(G)$ is the maximum cardinality of a maximal independent set of G. For example, the tree T in Fig. 1.7 has maximal independent sets of three sizes: $\{v_1, v_2, v_3, v_6, v_7\}$, $\{v_1, v_2, v_3, v_5\}$, and $\{v_4, v_6, v_7\}$. Thus, $i(T) = 3$ and $\alpha(T) = 5$. And T has minimal dominating sets of several different cardinalities: $\{v_4, v_5\}$, $\{v_4, v_6, v_7\}$, $\{v_1, v_2, v_3, v_5\}$, and $\{v_1, v_2, v_3, v_6, v_7\}$. Hence, $\gamma(T) = 2$ and $\Gamma(T) = 5$.

It is shown in Chap. 3 of [16] that to prove a dominating set S is minimal dominating, it suffices to show that $S \setminus \{v\}$ is not a dominating set for all $v \in S$. Further, to prove that an independent set S is maximal independent, it suffices to show that $S \cup \{v\}$ is not an independent set for all $v \in V \setminus S$. Stated in other words, an independent set S is maximal independent if and only if the following condition holds:

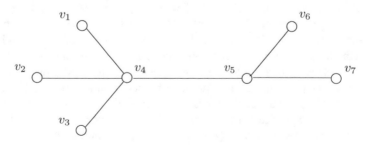

Fig. 1.7 Tree T

(1) For every vertex $u \in V \setminus S$, there is a vertex $v \in S$ such that u is adjacent to v.

Hedetniemi noticed that Condition (1) is precisely the definition of a dominating set, that is, the maximality condition for an independent set is the definition of a dominating set. Thus, the following holds, the necessity of which was first observed by Berge [3].

Proposition 1.3.1 ([3]) *An independent set S is maximal independent if and only if it is independent and dominating.*

Therefore, every maximal independent set is a dominating set. This is why the minimum cardinality of a maximal independent set is called the independent domination number $i(G)$.

Proposition 1.3.2 ([3]) *Every maximal independent set in a graph G is a minimal dominating set of G.*

Proof Let S be a maximal independent set in G. By Proposition 1.3.1, S is a dominating set. To see that S is, in fact, a minimal dominating set, it suffices to show that for every vertex $v \in S$, the set $S \setminus \{v\}$ is not a dominating set. Suppose, to the contrary, that there exists at least one vertex $v \in S$ for which $S \setminus \{v\}$ is a dominating set. Then v has at least one neighbor in $S \setminus \{v\}$, contradicting that S is an independent set. □

Proposition 1.3.2 implies the core part of the domination chain given in the following corollary

Corollary 1.3.3 *For any graph G, $\gamma(G) \leq i(G) \leq \alpha(G) \leq \Gamma(G)$.*

To complete the chain, we need additional definitions. The *open neighborhood* of a vertex v is the set $N(v) = \{u \in V \mid uv \in E(G)\}$, and the *closed neighborhood of v* is $N[v] = \{v\} \cup N(v)$. For a set of vertices $S \subseteq V$ and a vertex v belonging to the set S, the *S-private neighborhood* of v is defined by $pn[v, S] = \{w \in V \mid N[w] \cap S = \{v\}\}$, and a vertex of $pn[v, S]$ is called a *private neighbor* of v (with respect to S).

As first defined by Cockayne, Hedetniemi, and Miller [12], a vertex set S is an *irredundant set* of G if for every vertex $v \in S$, $pn[v, S] \neq \emptyset$, that is, every vertex $v \in S$ has at least one private neighbor. A set S is *maximal irredundant* if no proper superset of S is irredundant. The *irredundance number* $ir(G)$ is the minimum cardinality of a maximal irredundant set of G; while the *upper irredundance number* $IR(G)$ is the maximum cardinality of a maximal irredundant set of G. For example, consider the graph G in Fig. 1.8. Each of the sets $S = \{v_1, v_4\}$ and $S' = \{v_1, v_2, v_3\}$ is a maximal irredundant set. Note that for S, $pn[v_1, S] = \{v_2, v_3\}$, $pn[v_4, S] = \{v_5, v_6\}$; while for S', $pn[v_1, S'] = \{v_4\}$, $pn[v_2, S'] = \{v_5\}$, and $pn[v_3, S'] = \{v_6\}$. It can be shown that $ir(G) = 2$ and $IR(G) = 3$.

We note that the domination number of the graph G in Fig. 1.8 is also 2, so $ir(G) = \gamma(G) = 2$. Slater provided one of the first known examples of a graph having irredundance number strictly less than its domination number. This graph, known as

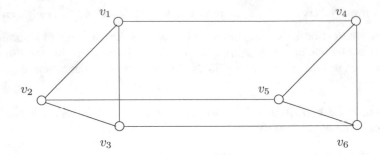

Fig. 1.8 Graph G

Fig. 1.9 The Slater H graph

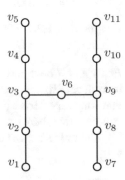

the Slater graph or the H graph, is illustrated in Fig. 1.9. The set $\{v_2, v_3, v_8, v_9\}$ is a minimum maximal irredundant set for H and the set $\{v_2, v_4, v_6, v_8, v_{10}\}$ is a minimum dominating set, and so $4 = ir(H) < \gamma(H) = 5$.

Recall what it means for a dominating set S to be minimal. If for any vertex $v \in S$, the set $S \setminus \{v\}$ is a dominating set, then S is not minimal. It follows that a dominating set S is a minimal dominating set if and only if:

(2) For every vertex $v \in S$, $pn[v, S] \neq \emptyset$, that is, every vertex $v \in S$ has at least one private neighbor.

Thus, the minimality condition for a dominating set is the definition of an irredundant set. This was first observed by Cockayne, Hedetniemi, and Miller [12] as follows.

Proposition 1.3.4 ([12]) *A dominating set S is a minimal dominating set if and only if it is dominating and irredundant.*

It is shown in [16] that an irredundant set is maximal irredundant if and only if $S \cup \{v\}$ is not an irredundant set for all $v \in V \setminus S$. Thus, an irredundant set S is maximal irredundant if and only if the following condition holds:

(3) For every vertex $v \in V \setminus S$, there exists a vertex $w \in S \cup \{v\}$ for which $pn[w, S \cup \{v\}] = \emptyset$.

The next result was first observed by Bollobás and Cockayne [5].

Proposition 1.3.5 ([5]) *Every minimal dominating set in a graph G is a maximal irredundant set of G.*

Proof By Proposition 1.3.4, every minimal dominating set S is an irredundant set, so all that remains is to show that S is maximal irredundant. Suppose it is not. It follows from our previous discussion that there exists a vertex $v \in V \setminus S$ for which $S \cup \{v\}$ is irredundant. Hence, $pn[v, S \cup \{v\}] \neq \emptyset$, that is, v has at least one private neighbor, say w, with respect to $S \cup \{v\}$. Thus, w is not adjacent to a vertex in S. Since S is a dominating set, v has a neighbor in S, and so $w \neq v$. Hence, $w \in V \setminus (S \cup \{v\})$. But then w is not dominated by S, contradicting the assumption that S is a dominating set. $\square$

Since every minimal dominating set is a maximal irredundant set, we now have the entire domination chain.

Theorem 1.3.6 (Cockayne, Hedetniemi, and Miller [12]) *For any graph G,*

$$ir(G) \leq \gamma(G) \leq i(G) \leq \alpha(G) \leq \Gamma(G) \leq IR(G).$$

Since being launched by Cockayne, Hedetniemi, and Miller [12] in 1978, the domination chain of Theorem 1.3.6 has been the focus of several hundred research papers and continues to intrigue researchers. Among the questions prompted by the domination chain are:

1. Given an integer sequence, $1 \leq a \leq b \leq c \leq d \leq e \leq f$, does there exist a graph G for which $1 \leq ir(G) = a \leq \gamma(G) = b \leq i(G) = c \leq \alpha(G) = d \leq \Gamma(G) = e \leq IR(G) = f$? If such a graph G does exist, then (a, b, c, d, e, f) is called a *domination sequence*.
2. Under what conditions are any pair of the parameters in the domination chain equal?
3. Are there variants of the basic independence-domination-irredundance parameters in the domination chain that satisfy a similar inequality chain?
4. Are there other graph parameters whose values are related to those in the domination chain? In particular, are there graph parameters whose values lie between two parameters in the domination chain?

Considering some examples of trees, we note that for the star $K_{1,n-1}$, $ir(K_{1,n-1}) = \gamma(K_{1,n-1}) = i(K_{1,n-1}) = 1$, while $\alpha(K_{1,n-1}) = \Gamma(K_{1,n-1}) = IR(K_{1,n-1}) = n - 1$. A *double star* $S(r, s)$ for $1 \leq r \leq s$ is a tree with exactly two (adjacent) vertices that are not leaves, one of which has r leaf neighbors and the other s leaf neighbors. Then $ir(S(r, s)) = \gamma(S(r, s)) = 2 \leq i(S(r, s)) = r + 1 \leq s + 1 = \alpha(S(r, s)) = \Gamma(S(r, s)) = IR(S(r, s))$, and the Slater graph H in Fig. 1.9 has $ir(T) = 4 < 5 = \gamma(T) = i(T) < 6 = \alpha(T) = \Gamma(T) = IR(T)$. Cockayne, Favaron, Payan, and Thomason [9] showed that the three upper parameters of the domination chain are equal for trees. Hence, the domination chain for trees can be stated as follows.

Theorem 1.3.7 *If T is a tree, then*

$$ir(T) \leq \gamma(T) \leq i(T) \leq \alpha(T) = \Gamma(T) = IR(T).$$

It is possible to obtain inequality chains similar to the domination chain starting from a suitable seed property. In conclusion, we mention two such chains that have been generated by Hedetniemi and his coauthors. Starting from the vertex cover number, Arumugam, Hedetniemi, Hedetniemi, Sathikala, and Sudha [1] defined the relevant graph parameters and developed the "covering chain" of inequalities using maximality and minimality conditions. Interestingly, they proved that this covering chain is, in a sense, the dual of the domination chain. For another example, Hedetniemi and his coauthors in [7] used the Roman domination number as the seed for the chain. They defined the Roman independence number, the upper Roman domination number, and the upper and lower Roman irredundance numbers to develop a Roman domination chain paralleling the domination chain. A section on Roman domination is given in Chap. 4.

References

1. S. Arumugam, S.T. Hedetniemi, S.M. Hedetniemi, L. Sathikala, S. Sudha, The covering chain of a graph. Util. Math. **98**, 183–196 (2015)
2. W.W. Rouse Ball, *Mathematical Recreations and Essays*, revision by H.S.M. Coxeter of the original 1892; *Minimum Pieces Problem*, 3rd edn. (Macmillan, London, 1939)
3. C. Berge, *Theory of Graphs and Its Applications* (Methuen, London, 1962)
4. W. Bird, *Computational Methods for Domination Problems*. Doctoral Dissertation, University of Victoria, 2017
5. B. Bollobás, E.J. Cockayne, Graph theoretic parameters concerning domination, independence and irredundance. J. Graph Theory **3**, 241–250 (1979)
6. S. Bozóki, P. Gál, I. Marosi W.D. Weakley, Domination of the rectangular queens graph, submitted
7. M. Chellali, T.W. Haynes, S.M. Hedetniemi, S.T. Hedetniemi, A. McRae, A Roman domination chain. Graphs Comb. **32**, 79–92 (2016)
8. E.J. Cockayne, R.M. Dawes, S.T. Hedetniemi, Total domination in graphs. Networks **10**, 211–219 (1980)
9. E.J. Cockayne, O. Favaron, C. Payan, A. Thomason, Contributions to the theory of domination, independence, and irredundance in graphs. Discret. Math. **33**, 249–258 (1981)
10. E.J. Cockayne, J.H. Hattingh, S.M. Hedetniemi, S.T. Hedetniemi, A.A. McRae, Using maximality and minimality conditions to construct inequality chains. Discret. Math. **176**, 43–61 (1997)
11. E.J. Cockayne, S.T. Hedetniemi, On the diagonal queens domination problem. J. Comb. Theory Ser. A **42**, 137–139 (1986)
12. E.J. Cockayne, S.T. Hedetniemi, D.J. Miller, Properties of hereditary hypergraphs and middle graphs. Canad. Math. Bull. **21**, 461–468 (1978)
13. C.F. De Jaenisch, *Applications de l'Analyse Mathematique an Jeu des Echecs* (Petrograd, 1863)
14. D. Finozhenok, W.D. Weakley, An improved lower bound for domination numbers of the queen's graph. Australas. J. Comb. **37**, 295–300 (2007)
15. G.H. Fricke, S.M. Hedetniemi, S.T. Hedetniemi, A. McRae, C.K. Wallis, M.S. Jacobson, H.W. Martin, W.D. Weakley, Combinatorial problems on chessboards: a brief survey. *Graph Theory,*

Combinatorics, and Algorithms, vol. 1, 2 (Kalamazoo, 1992), pp. 507–528; (Wiley-Interscience Publication, Wiley, New York, 1995)

16. T.W. Haynes, S.T. Hedetniemi, P.J. Slater, *Fundamentals of Domination in Graphs* (Marcel Dekker, New York, 1998)
17. S.M. Hedetniemi, S.T. Hedetniemi, R. Reynolds, Combinatorial problems on chessboards. II, *Domination in Graphs*, Monographs and Textbooks in Pure and Applied Mathematics, vol. 209 (Dekker, New York, 1998), pp. 133–162
18. S.T. Hedetniemi, My top 10 graph theory conjectures and open problems, in *Graph Theory, Favorite Conjectures and Open Problems, Volume 1*, ed. by R. Gera, S.T. Hedetniemi, C. Larson (Springer, Berlin, 2018), pp. 177–281
19. D. König, *Theorie der endlichen und unendlichen graphen* (Akademische Verlagsgesellschaft M.B.H. Leipzig, 1936) and (Chelsea Publishing, New York, 1950)
20. P.R.J. Östergård, W.D. Weakley, Values of domination numbers of the queen's graph. Electron. J. Comb. **8** (2001), Research Paper 29, 19 pp
21. A. Sinko, P.J. Slater, Queens domination using border squares and (A, B)-restricted domination. Discret. Math. **308**, 4822–4828 (2008)
22. W.D. Weakley, A lower bound for domination numbers of the queen's graph. J. Comb. Math. Comb. Comput. **43**, 231–254 (2002)
23. W.D. Weakley, Upper bounds for domination numbers of the queen's graph. Discret. Math. **242**, 229–243 (2002)
24. W.D. Weakley, Domination in the queen's graph, *Graph Theory, Combinatorics, and Algorithms*, vol. 1, 2 (Kalamazoo, 1992), pp. 1223–1232; (Wiley-Interscience Publication, Wiley, New York, 1995)
25. W.D. Weakley, Queens around the world in twenty-five years, in *Graph Theory, Favorite Conjectures and Open Problems, Volume 2*, ed. by R. Gera, T.W. Haynes, S.T. Hedetniemi (Springer, Berlin, 2018), pp. 43–54

Chapter 2
Key Domination Parameters

In this chapter, two key domination parameters different from the standard domination number are discussed. The first, the independent domination number, was introduced by Stephen Hedetniemi and his coauthor Ernie Cockayne in 1974. The second, the total domination number, was introduced by Hedetniemi together with his coauthors Cockayne and Dawes in 1980.

2.1 Introduction

As we pointed out in Chap. 1, Stephen Hedetniemi is one of the pioneers in the area of domination in graphs. We recall that a set S of vertices in a graph G is a *dominating set* of G if every vertex in $V(G) \backslash S$ is adjacent to a vertex in S and the *domination number* $\gamma(G)$ of G is the minimum cardinality of a dominating set of G. Further, the *upper domination number* $\Gamma(G)$ of G is the maximum cardinality of a minimal dominating set of G. A set S of vertices is independent if no two vertices in S are adjacent, and the maximum cardinality of an independent set in a graph G is its *independence number*, denoted $\alpha(G)$.

In this chapter, we study two variations of domination pioneered by Stephen Hedetniemi, namely independent domination and total domination.

2.2 Independent Domination in Graphs

Berge [3] was the first to observe that an independent set is a maximal independent set if and only if it is independent and dominating. Therefore, every maximal independent set is a dominating set. Early in the 1970s, Stephen Hedetniemi understood that this fundamental property of an independent set had an important place in the theory of domination in graphs, and he began a study of dominating sets that have this

G. Chartrand et al., *From Domination to Coloring*, SpringerBriefs in Mathematics, https://doi.org/10.1007/978-3-030-31110-0_2

additional property of being independent. More formally, an *independent dominating set* of a graph G is a set S of vertices of G that is both independent and dominating. Equivalently, no two vertices of S are adjacent, and every other vertex is adjacent to at least one vertex of S. The *independent domination number* $i(G)$ of G is the minimum cardinality of an independent dominating set of G.

Berge [3] was also the first to observe that every maximal independent set in a graph G is a minimal dominating set of G. Hence we have the following inequality chain.

Theorem 2.2.1 ([3]) *For every graph* G, $\gamma(G) \leq i(G) \leq \alpha(G) \leq \Gamma(G)$.

The independent domination number of a graph is therefore sandwiched between its domination number and its independence number. The inequality chain in Theorem 2.2.1 is part of the canonical domination chain which was first observed by Cockayne et al. [12] in 1978, and generated considerable interest in the domination theory community with several hundred papers emanating from it.

The domination chain stated in Theorem 2.2.1 suggests the following question. Given integers s_1, s_2, s_3, s_4, does there exist a graph G for which $\gamma(G) = s_1$, $i(G) = s_2$, $\alpha(G) = s_3$, and $\gamma(G) = s_4$? If such a graph G exists, then the sequence (s_1, s_2, s_3, s_4) of integers is called a *domination sequence*. The following 1993 result, due to Cockayne and Mynhardt [11], characterizes domination sequences.

Theorem 2.2.2 ([11]) *A sequence* (s_1, s_2, s_3, s_4) *of integers is a domination sequence if and only if the following three conditions hold.*

(a) $1 \leq s_1 \leq s_2 \leq s_3 \leq s_4$.
(b) $s_1 = 1$ *implies that* $s_2 = 1$.
(c) $s_3 = 1$ *implies that* $s_4 = 1$.

A section on the domination chain is also given in Chap. 1. A detailed discussion on the domination chain can be found in Chap. 3 in [18].

At the Fifth Southeastern Conference on Combinatorics, Graph Theory and Computing held at Florida Atlantic University in 1974, Cockayne and Hedetniemi [7] started their prolific series of joint papers on a systematic study of domination in graphs. In this paper they defined the independence graph $I(G)$ of a graph G as the graph whose vertices can be put in a one-to-one correspondence with the independent sets of vertices of G and where two vertices of $I(G)$ are adjacent if the corresponding independent sets of vertices of G overlap. Influenced by Hedetniemi's earlier work on colorings in graphs, they showed that the chromatic number of G equals the independent domination number of the graph $I(G)$. Further, they related the independent domination number to a variety of other graph parameters including thickness, arboricity, and the Hamiltonian completion number.

However, it was their 1976 paper [10] entitled "Towards a theory of domination in graphs", written several years earlier, that proved to be the main launching pad for a systematic study of domination in graphs. This paper has to date been cited well over 600 times. In this paper, they showed that the theory of domination resembles the theory of colorings of graphs in the following sense. A coloring of a graph involves independent sets of vertices (called the color classes of the coloring) which is a hereditary property in that it is satisfied by every subset of an independent set, while domination is an expanding property as every superset of a dominating set is one. They coined the term *indominable graphs* to describe those graphs whose vertex set can be partitioned completely into independent dominating sets. For example, a 5-cycle is not an indominable graph since every independent dominating set in the graph has size exactly 2.

A fundamental property of dominating sets is that every graph with no isolated vertex contains two vertex disjoint dominating sets. However, it is not true that every graph with no isolated vertex contains two vertex disjoint independent dominating sets. Recall that the *corona* cor(G) (sometimes denoted $G \circ P_1$ in the literature) is the graph obtained from G by adding a pendant edge at each vertex of G. The corona cor(C_3) of a 3-cycle, for example, does not have two vertex disjoint dominating sets. The *idomatic number* of a graph G, which we denote by idom(G), is the maximum number of vertex disjoint independent dominating sets in G. This terminology was introduced by Zelinka [28], but the parameter was originally defined by Cockayne and Hedetniemi in their 1976 paper [9] on disjoint independent dominating sets in graphs. In this paper, Cockayne and Hedetniemi studied the following conjecture posed by Berge.

Conjecture 2.2.3 (Berge) *Every k-regular graph where $k \geq 1$ has two disjoint independent dominating sets.*

Conjecture 2.2.3 is easy to verify for $k = 1$ and $k = 2$. For $k = 3$ and $k = 4$, the truth of the conjecture is attributed to Berge in [9] (although there is no available published manuscript proving the conjecture in this case). When $k = 3$ this implies that every cubic graph has two disjoint independent dominating sets. This result has recently been strengthened slightly by Goddard and Henning [17].

Theorem 2.2.4 ([17]) *If G is a graph with minimum degree at least 2 and maximum degree at most 3, then G has two disjoint independent dominating sets.*

The main contribution of Cockayne and Hedetniemi in [9] is to prove Berge's Conjecture 2.2.3 for large k, namely when $k \geq n - 7$ where n is the order of the k-regular graph.

As observed in Theorem 2.2.2, the difference between the independent domination number and the domination number can be made arbitrarily large. In 1962 Ore [24] was the first to observe that if G is a graph with no isolated vertex, then $\gamma(G) \leq \frac{1}{2}n$. However unlike the domination number, there is no constant $C < 1$ such that for every graph G with no isolated vertex, the bound $i(G) \leq C \cdot n$ holds, as first observed by Favaron [14] in 1988.

Theorem 2.2.5 ([14]) *If G is a graph of order n with no isolated vertex, then $i(G) \leq n + 2 - 2\sqrt{n}$.*

That the bound of Theorem 2.2.5 is tight may be seen as follows. The *generalized corona* $\mathrm{cor}(G, r)$ of a graph G is the graph obtained from G by adding r pendant edges to each vertex of G. In particular, if $r = 1$, then we note that $\mathrm{cor}(G, r)$ is the corona $\mathrm{cor}(G)$ defined earlier. For $k \geq 2$ if we take $G = \mathrm{cor}(K_k, k - 1)$, then G has order $n = k^2$ and $i(G) = (k - 1)^2 + 1 = n + 2 - 2\sqrt{n}$.

Given a connected graph G with arbitrary minimum degree $\delta \geq 2$, a tight upper bound (that holds for graphs of arbitrarily large order) on $\gamma(G)$ has yet to be determined, even for the special case when $\delta = 3$. However, this is not the case for the independent domination number. Favaron [14] conjectured such an upper bound on $i(G)$ as a function on n and δ. This conjecture was proven in general by Sun and Wang [25] in 1999.

Theorem 2.2.6 ([25]) *If G is a graph of order n with minimum degree $\delta \geq 2$, then $i(G) \leq n + 2\delta - 2\sqrt{\delta n}$.*

Earlier, Favaron [14] showed that for every positive integer δ, the bound in Theorem 2.2.6 is attained for infinitely many graphs. Thus, the independent domination number behaves very differently from the classical domination number. For a more detailed discussion on the independent domination number of a graph, we refer the reader to a survey on independent domination in graphs by Goddard and Henning [15].

2.3   Total Domination in Graphs

In Sect. 2.2, we remarked that Stephen Hedetniemi understood that the fundamental properties of an independent set have an important place in the theory of domination in graphs, and thus he began a study of dominating sets that have this additional property of being independent.

In this section, we consider a property of a dominating set that is antipodal to that of independence, namely that every dominating set induces a subgraph with no isolated vertex. Stephen Hedetniemi understood that this property of a dominating set had an important place in the theory of domination in graphs, and he began a study of dominating sets having the additional property that every vertex in the dominating set has at least one neighbor in the set. The rest is now history, and this concept of domination plays a fundamental role in the theory of domination in graphs.

More formally, a *total dominating set* of a graph G is a set S of vertices of G such that every vertex has a neighbor in S. Equivalently, every vertex is adjacent to at least one vertex of S (different from itself). The *total domination number* $\gamma_t(G)$ of G is the minimum cardinality of a total dominating set of G. Hedetniemi began his study of total dominating sets in graphs with his seminal 1980 paper [8] with Cockayne and Dawes. This paper, which birthed the concept of total domination in graphs, has to date been cited well over 600 times.

In [8], Cockayne, Dawes, and Hedetniemi established fundamental properties of a minimal total dominating set in a graph. In order to state these properties, we need some additional terminology. For a graph G and set S of vertices in G, the *S-private neighborhood* of a vertex $v \in S$ is defined by $\text{pn}(v, S) = \{w \in V(G) \mid N_G(w) \cap S = \{v\}\}$ and a vertex in $\text{pn}(v, S)$ is called an S-private neighbor of v in G. The sets $\text{ipn}(v, S) = \text{pn}(v, S) \cap S$ and $\text{epn}(v, S) = \text{pn}(v, S) \backslash S$ consist of the S-private neighbors of v that belong to the set S and do not belong to the set S, respectively. The set $\text{ipn}(v, S)$ is the S-internal private neighborhood of v, while the set $\text{epn}(v, S)$ is the S-external private neighborhood of v. Further, a vertex in $\text{ipn}(v, S)$ (respectively, $\text{epn}(v, S)$) is called an S-internal (respectively, S-external) private neighbor of v. We note that $\text{pn}(v, S) = \text{ipn}(v, S) \cup \text{epn}(v, S)$. We are now in a position to state the fundamental property of minimal total dominating sets established in [8].

Theorem 2.3.1 ([8]) *Let S be a total dominating set in a graph G. The set S is a minimal total dominating set in G if and only if $|\text{epn}(v, S)| \geq 1$ or $|\text{ipn}(v, S)| \geq 1$ for each $v \in S$.*

We remark that the following stronger property of a minimum TD-set in a graph is established in [20].

Theorem 2.3.2 ([20]) *If G is a connected graph of order $n \geq 3$ and $G \not\cong K_n$, then G has a minimum total dominating set S such that every vertex $v \in S$ satisfies $|\text{epn}(v, S)| \geq 1$ or is adjacent to a vertex v' of degree 1 in $G[S]$ satisfying $|\text{epn}(v', S)| \geq 1$.*

Using the properties of a minimal total dominating set in a graph in Theorem 2.3.1, Cockayne et al. [8] proceeded to establish a tight upper bound on the total domination number of a connected graph in terms of its order as follows.

Theorem 2.3.3 ([8]) *If G is a connected graph of order $n \geq 3$, then $\gamma_t(G) \leq \frac{2}{3}n$.*

Proof Let G be a connected graph of order $n \geq 3$. If $G = K_n$, then $\gamma_t(G) = 2 \leq \frac{2}{3}n$. Hence, we may assume that $G \neq K_n$. Applying Theorem 2.3.2 to the graph G, we note that there exists a minimum total dominating S in G satisfying the statement of that theorem. Let $A = \{v \in S \mid \text{epn}(v, S) = \emptyset\}$ and let $B = S \backslash A$. By Theorem 2.3.2, each vertex $v \in A$ has a neighbor in B that is adjacent to v but to no other vertex of S. Hence, $|A| \leq |B|$ and $|S| = |A| + |B| \leq 2|B|$, and so $|B| \geq |S|/2$. Let C be the set of all S-external private neighbors. We note that $C \subseteq V(G) \backslash S$. Further, since $\text{epn}(v, S) \geq 1$ for each vertex $v \in B$, we note that $|C| \geq |B|$. Hence, $n - |S| = |V(G) \backslash S| \geq |C| \geq |B| \geq |S|/2$, and so $\gamma_t(G) = |S| \leq \frac{2}{3}n$. $\square$

Brigham et al. [4] characterized the infinite family of connected graphs that achieve equality in the Cockayne-Dawes-Hedetniemi bound of $\frac{2}{3}n$; that is, they characterized the connected graphs of order at least 3 with total domination number exactly two-thirds their order. For this purpose, we define the *2-corona* $H \circ P_2$ of a graph H to be the graph of order $3|V(H)|$ obtained from H by attaching a path of length 2 to each vertex of H so that the resulting paths are vertex-disjoint. The proof, which we omit here, of the characterization in [4] follows relatively easily from the properties established by Cockayne et al. [8] of a minimal total dominating set in a graph in Theorem 2.3.1.

Theorem 2.3.4 ([4]) *Let G be a connected graph of order $n \geq 3$. Then $\gamma_t(G) = \frac{2}{3}n$ if and only if G is C_3, C_6, or $F \circ P_2$ for some connected graph F.*

The upper bound in Theorem 2.3.3 cannot be improved if we simply restrict the minimum degree to be 2, as may be seen by taking G to be a graph consisting of a disjoint union of 3-cycles and 6-cycles. In this case, $\gamma_t(G) = \frac{2}{3}n$. However, if we impose the additional restriction that G is connected, then the $\frac{2}{3}n$-bound can be improved to a $\frac{3}{7}n$-bound, except for six small exceptional graphs of orders at most 10, as shown in [20].

We summarize the known upper bounds on the total domination number of a graph G in terms of its order n and minimum degree δ in Table 2.1.

The infinite family of connected graphs G of order $n > 14$ with $\delta(G) \geq 2$ satisfying $\gamma_t(G) = \frac{4}{7}n$ are characterized in [20], and the infinite family of connected graphs G of order n with $\delta(G) \geq 3$ satisfying $\gamma_t(G) = \frac{1}{2}n$ are characterized in [22]. It is shown in [23] that the bipartite complement of the Heawood graph is the unique connected graph G of order n with $\delta(G) \geq 4$ satisfying $\gamma_t(G) = \frac{3}{7}n$.

However, it is unlikely that the upper bound of $(\frac{4}{11} + \frac{1}{72})n$ shown in Table 2.1 when $\delta(G) \geq 5$ is achievable. Indeed, in this case Thomasse and Yeo [26] posed the conjecture that if G is a graph of order n with $\delta(G) \geq 5$, then $\gamma_t(G) \leq \frac{4}{11}n$. If this conjecture is true, then the bound is achievable. For example, the graph G_{22}, shown in Fig. 2.1, has order $n = 22$, minimum degree $\delta(G_{22}) = 5$, and $\gamma_t(G_{22}) = 8 = \frac{4}{11}n$.

In their seminal 1980 paper [8], Cockayne, Dawes, and Hedetniemi also introduced and first studied the total domatic number of a graph, which has subsequently attracted a great deal of interest. The *total domatic number* tdom(G) of a graph G is the maximum number of disjoint total dominating sets in G. The parameter tdom(G)

Table 2.1 Upper bounds on the total domination number of a graph G

$\delta(G) \geq 1$	$\Rightarrow \gamma_t(G) \leq \frac{2}{3}n$	if $n \geq 3$ and G is connected	([8])
$\delta(G) \geq 2$	$\Rightarrow \gamma_t(G) \leq \frac{4}{7}n$	if $n \geq 11$ and G is connected	([20])
$\delta(G) \geq 3$	$\Rightarrow \gamma_t(G) \leq \frac{1}{2}n$		([2, 6, 27])
$\delta(G) \geq 4$	$\Rightarrow \gamma_t(G) \leq \frac{3}{7}n$		([26])
$\delta(G) \geq 5$	$\Rightarrow \gamma_t(G) \leq \left(\frac{4}{11} + \frac{1}{72}\right)n$		([13])
Any $\delta(G) \geq 1$	$\Rightarrow \gamma_t(G) \leq \left(\frac{1+\ln\delta}{\delta}\right)n$		([21])

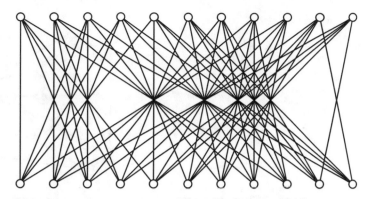

Fig. 2.1 The graph G_{22}

is equivalent to the maximum number of colors in a so-called *coupon coloring* of G which is a (not necessarily proper) coloring of the vertices of G where every color is thoroughly dispersed, that is, appears in every open neighborhood. In their 2015 paper Chen et al. [5] called this the *coupon coloring problem*.

The total domatic number is now well studied. In 1989 Zelinka [29] showed that there are graphs with arbitrarily large minimum degree without two disjoint total dominating sets, implying that such graphs G have total domatic number $\text{tdom}(G) = 1$. Heggernes and Telle [19] showed that the decision problem to decide for a given graph G if $\text{tdom}(G) \geq 2$ is NP-complete, even for bipartite graphs. In contrast, several researchers, such as Aram et al. [1], studied $\text{tdom}(G)$ for a k-regular graph G; in particular, Chen et al. [5] showed that such graphs have total domatic number at least $(1 - o(1))\frac{k}{\ln k}$.

Goddard and Henning [16] showed that the total domatic number of a planar graph is at most 4.

Theorem 2.3.5 ([16]) *If G is a planar graph, then* $\text{tdom}(G) \leq 4$.

There do exist planar graphs G with $\text{tdom}(G) = 4$. As shown in [16], if we take the truncated tetrahedron and add a vertex inside each hexagonal face that is joined to all vertices on the boundary, then the resulting planar graph has total domatic number equal to 4. Illustrated below (see Fig. 2.2) is a spanning subgraph of this graph that still has four disjoint total dominating sets: the vertices labelled i form a total dominating set for each $i \subset [4]$.

Goddard and Henning [16] also studied the total domatic number of graphs on other surfaces. For example, they showed that the total domatic number of a toroidal graph is at most 5.

Theorem 2.3.6 ([16]) *If G is a toroidal graph, then* $\text{tdom}(G) \leq 5$.

As remarked in [16], there do exist toroidal graphs G with $\text{tdom}(G) = 5$. Such an example, given by Goddard and Henning in [16], is illustrated in Fig. 2.3, where

Fig. 2.2 A planar graph G
with $\mathrm{tdom}(G) = 4$

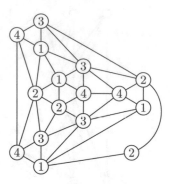

Fig. 2.3 A toroidal graph G
with $\mathrm{tdom}(G) = 5$

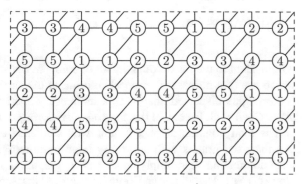

the top and bottom dotted lines should be identified and similarly with the left and right dotted lines. In this illustration, the vertices labelled i form a total dominating set of G for each $i \in [5]$.

We close this section with a few remarks. A total dominating set in a graph is in some sense more fundamental than a dominating set, in that every vertex is now required to be dominated by some other vertex. Much of the recent interest in total domination in graphs arises from the fact that total domination in graphs can be translated to the problem of finding transversals in hypergraphs. The open neighborhood hypergraph, abbreviated ONH, of a graph G is the hypergraph H_G with the same vertex set as G and whose edge set consists of the open neighborhoods on vertices of $V(G)$ in G. Thus, for each vertex x in G, the set $N_G(x)$ consisting of the neighbors of x in G is an edge in H_G. If a graph G has order n, then its ONH has order n and size n (noting there is a one-to-one correspondence between the edges of the ONH of G and the n open neighborhoods of vertices in G, one associated with each of its n vertices). The transversal number of the ONH of a graph is precisely the total domination number of the graph, where a transversal in a hypergraph is a set of vertices intersecting every edge of the hypergraph.

Observation 2.3.7 *If G is a graph with no isolated vertex and H_G is the ONH of G, then $\gamma_t(G) = \tau(H_G)$.*

As observed in [23], the main advantage of considering hypergraphs rather than graphs is that the structure is easier to handle and we can move to hypergraphs which are not open neighborhood graphs of any graph, giving much more flexibility than just using graphs. The idea of using transversals in hypergraphs to obtain results on total domination in graphs was introduced by Thomassé and Yeo [26] in 2003. Up to that time, the transition from total domination in graphs to transversals in hypergraphs seemed to pass by unnoticed. Subsequent to the Thomassé-Yeo paper, several important results on total domination in graphs have been obtained using transversals in hypergraphs.

The graph theory community is greatly indebted to Stephen Hedetniemi for birthing the concept of total domination in graphs, and for his depth of insight and understanding that this property of a dominating set in which every vertex of the graph has at least one neighbor in the set has an important place in the theory of domination in graphs. For a more detailed discussion on the total domination number of a graph, we refer the reader to the book by Henning and Yeo [23].

References

1. H. Aram, S.M. Sheikholeslami, L. Volkmann, On the total domatic number of regular graphs. Trans. Comb. **1**, 45–51 (2012)
2. D. Archdeacon, J. Ellis-Monaghan, D. Fischer, D. Froncek, P.C.B. Lam, S. Seager, B. Wei, R. Yuster, Some remarks on domination. J. Graph Theory **46**, 207–210 (2004)
3. C. Berge, *Theory of Graphs and its Applications* (Methuen, London, 1962)
4. R.C. Brigham, J.R. Carrington, R.P. Vitray, Connected graphs with maximum total domination number. J. Combin. Comput. Combin. Math. **34**, 81–96 (2000)
5. B. Chen, J.H. Kim, M. Tait, J. Verstraete, On coupon colorings of graphs. Discret. Appl. Math. **193**, 94–101 (2015)
6. V. Chvátal, C. McDiarmid, Small transversals in hypergraphs. Combinatorica **12**, 19–26 (1992)
7. E.J. Cockayne, S.T. Hedetniemi, Independence graphs, in *Proceedings of the Fifth Southeastern Conference on Combinatorics, Graph Theory and Computing* (Florida Atlantic University, Boca Raton, FL, 1974), pp. 471–491. Congress. Numer., No. X, *Utilitas Math.*, Winnipeg, Man. 1974
8. E.J. Cockayne, R.M. Dawes, S.T. Hedetniemi, Total domination in graphs. Networks **10**(3), 211–219 (1977)
9. E.J. Cockayne, S.T. Hedetniemi, Disjoint independent dominating sets in graphs. Discret. Math. **15**(3), 213–222 (1976)
10. E.J. Cockayne, S.T. Hedetniemi, Towards a theory of domination in graphs. Networks **7**(3), 247–261 (1977)
11. E.J. Cockayne, C.M. Mynhardt, The sequence of upper and lower domination, independence and irredundance numbers of a graph. Discret. Math. **122**, 89–102 (1993)
12. E.J. Cockayne, S.T. Hedetniemi, D.J. Miller, Properties of hereditary hypergraphs and middle graphs. Canad. Math. Bull. **21**, 461–468 (1978)
13. A. Eustis, M.A. Henning, A. Yeo, Independence in 5-uniform hypergraphs. Discret. Math. **339**, 1004–1027 (2016)
14. O. Favaron, Two relations between the parameters of independence and irredundance. Discret. Math. **70**, 17–20 (1988)
15. W. Goddard, M.A. Henning, Independent domination in graphs: A survey and recent results. Discret. Math. **313**, 839–854 (2013)

16. W. Goddard, M.A. Henning, Thoroughly dispersed colorings. J. Graph Theory **88**(1), 174–191 (2018)
17. W. Goddard, M.A. Henning, Acyclic total dominating sets in cubic graphs. To appear in Appl. Anal. Discr. Math.
18. T.W. Haynes, S.T. Hedetniemi, P.J. Slater, *Fundamentals of Domination in Graphs* (Marcel Dekker Inc., New York, 1998)
19. P. Heggernes, J.A. Telle, Partitioning graphs into generalized dominating sets. Nordic J. Comput. **5**, 128–142 (1998)
20. M.A. Henning, Graphs with large total domination number. J. Graph Theory **35**(1), 21–45 (2000)
21. M.A. Henning, A. Yeo, A transition from total domination in graphs to transversals in hypergraphs. Quaest. Math. **30**, 417–436 (2007)
22. M.A. Henning, A. Yeo, Hypergraphs with large transversal number and with edge sizes at least three. J. Graph Theory **59**, 326–348 (2008)
23. M.A. Henning, A. Yeo, *Total domination in graphs* (Springer Monographs in Mathematics, 2013). ISBN: 978-1-4614-6524-9 (Print) 978-1-4614-6525-6 (Online)
24. O. Ore, Theory of graphs. Am. Math. Soc. Transl. **38**, 206–212 (1962)
25. L. Sun, J. Wang, An upper bound for the independent domination number. J. Combin. Theory Ser. B **76**, 240–246 (1999)
26. S. Thomassé, A. Yeo, Total domination of graphs and small transversals of hypergraphs. Combinatorica **27**, 473–487 (2007)
27. Zs. Tuza, Covering all cliques of a graph. Discret. Math. **86**, 117–126 (1990)
28. B. Zelinka, Adomatic and idomatic numbers of graphs. Math. Slovaca **33**, 99–103 (1983)
29. B. Zelinka, Total domatic number and degrees of vertices of a graph. Math. Slovaca **39**, 7–11 (1989)

Chapter 3
Dominating Functions

In this chapter, we discuss dominating functions in graphs, a concept birthed by Stephen Hedetniemi in the mid 1980s.

3.1 Introduction

Stephen Hedetniemi understood that a fundamental property of a dominating set S in a graph G is that the characteristic function of the set S (that assigns to every vertex in the set S the value of 1 and to every vertex outside the set S the value 0) has the property that the sum of the function values assigned to any given vertex and its neighbors in G will always sum to at least 1. He would often refer to this property as the "dominating property" of a function. This motivated him to define a *dominating function* of a graph G as a function $f : V(G) \to \{0, 1\}$ such that for every vertex v of G, the function values of f summed over all vertices in the closed neighborhood $N_G[v]$ of v is at least 1. The *weight* $w(v)$ of a vertex v is its value $f(v)$ assigned to it under f. The *weight* $w(f)$ of f is the sum $\sum_{u \in V(G)} f(u)$ of the weights of the vertices in the graph G. For a set S of vertices of G, the *weight* of the set S is $w(S) = \sum_{v \in S} f(v)$. For notational convenience, we denote the weight $f(N_G[v])$ of the closed neighborhood of a vertex v simply by $f[v]$. We note that the domination number of G can be defined as

$$\gamma(G) = \min\{w(f) \mid f \text{ is a dominating function in } G\}.$$

A dominating function f is a *minimal dominating function* if there does not exist a dominating function $g : V(G) \to \{0, 1\}$ different from f for which $g(v) \le f(v)$ for every $v \in V(G)$. This is equivalent to saying that a dominating function f is minimal if for every vertex v with $f(v) = 1$, there exists a vertex $u \in N_G[v]$ satisfying

$f[u] = 1$. Thus, the *upper domination number* of G can be defined as

$$\Gamma(G) = \max\{w(f) \mid f \text{ is a minimal dominating function in } G\}.$$

In this chapter, we discuss the concept proposed by Hedetniemi of changing the allowable weights of the vertices, provided that the "dominating property" that the sum of the weights in the closed neighborhood of each vertex is at least 1 is preserved.

3.2   Fractional Dominating Functions

In 1984 Farber [10] introduced indirectly the concept of fractional domination in graphs. Reporting on results in [17] at the Eighteenth Southeastern Conference, Stephen Hedetniemi formally defined fractional domination as follows. For a graph $G = (V, E)$, a function $f : V \to [0, 1]$ is a *fractional dominating function* of G if $f[v] \geq 1$ for each $v \in V$. The *fractional domination number*, denoted $\gamma_f(G)$, and the *upper fractional domination number*, denoted $\Gamma_f(G)$, of G are defined by

$$\gamma_f(G) = \min\{w(f) \mid f \text{ is a minimal fractional dominating function for } G\}$$
$$\Gamma_f(G) = \max\{w(f) \mid f \text{ is a minimal fractional dominating function for } G\}$$

For example, the function f_1 that assigns the weights to the vertices of the Hajós graph H as illustrated in Fig. 3.1a is a minimal fractional dominating function of H of weight $w(f_1) = \frac{3}{2}$. Thus, $\gamma_f(H) \leq \frac{3}{2}$. In fact, $\gamma_f(H) = \frac{3}{2}$. For a positive integer k, the function f_2 that assigns the weights to the vertices of H as illustrated in Fig. 3.1b is a minimal fractional dominating function of H of weight $w(f_2) = 3(k + 1)/(2k)$. In particular, if $k = 1$, then the function f_2, illustrated in Fig. 3.1c, is a minimal fractional dominating function of H of weight $w(f_2) = 3$. Thus, $\Gamma_f(H) \geq 3$. In fact, $\Gamma_f(H) = 3$.

The domination chain in Theorem 2.2.1 can be extended by adding to it the fractional domination and the upper fractional domination numbers as follows.

Theorem 3.2.1 *For every graph* G, $\gamma_f(G) \leq \gamma(G) \leq i(G) \leq \alpha(G) \leq \Gamma(G) \leq \Gamma_f(G)$.

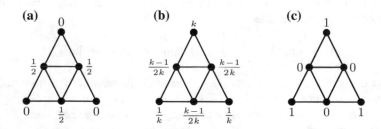

Fig. 3.1   Minimal fractional dominating functions of the Hajós graph H

For a more detailed discussion on this fractional version of domination, we refer the reader to the survey on fractional domination in graphs by Domke et al. [4].

3.3 Minus Dominating Functions

In the early 1990s, Stephen Hedetniemi put forward the idea of allowing negative weights in the mix. This resulted in the concept of minus domination. For a graph $G = (V, E)$, a function $f : V \rightarrow \{-1, 0, 1\}$ is a *minus dominating function* of G if $f[v] \geq 1$ for each $v \in V$. The *minus domination number*, denoted $\gamma^-(G)$, and the *upper minus domination number*, denoted $\Gamma^-(G)$, of G are defined by

$$\gamma^-(G) = \min \{w(f) \mid f \text{ is a minimal minus dominating function for } G\}$$
$$\Gamma^-(G) = \max \{w(f) \mid f \text{ is a minimal minus dominating function for } G\}.$$

For example, the function f_1 that assigns the weights to the vertices of the Hajós graph H as illustrated in Fig. 3.2a is a minimal minus dominating function of H of weight $w(f_1) = 0$. Thus, $\gamma^-(H) \leq 0$. In fact, $\gamma^-(H) = 0$. Moreover, the function f_2 that assigns the weights to the vertices of the Hajós graph H as illustrated in Fig. 3.2b is a minimal minus dominating function of H of weight $w(f_2) = 3$. Thus, $\Gamma^-(H) \geq 3$. In fact, $\Gamma^-(H) = 3$.

One of the applications for this variation of domination given by Hedetniemi is that by assigning the values $-1, 0$, or $+1$ to the vertices of a graph that models a network of people or organizations in which global decisions must be made (e.g., positive, negative, or neutral responses or preferences). As explained in [18], in such a context, "the minus domination number represents the minimum number of people whose positive votes can assure that all local groups of voters (represented by closed neighborhoods) have more positive than negative voters, even though the entire network may have far more people who vote negative than positive. By contrast, the upper minus domination number represents the greatest number of positive voters that may be required to offset a few negative voters, i.e., to insure that all local groups of voters have positive vote totals."

In [8], the authors showed that Γ and Γ^- are not comparable. Indeed, there exist graphs G and H such that $\Gamma(G) < \Gamma^-(G)$ and $\Gamma(H) > \Gamma^-(H)$. However, the char-

Fig. 3.2 The Hajós graph H

(a)

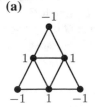

(b)

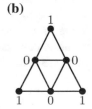

acteristic function of a minimum dominating set of a graph is a minimal minus dominating function, and the characteristic function of a maximum independent set of a
graph is a minimal minus dominating function, implying the following domination
chain.

Theorem 3.3.1 *For every graph G, $\gamma^-(G) \leq \gamma(G) \leq i(G) \leq \alpha(G) \leq \Gamma^-(G)$.*

The domination and the minus domination number of a tree are related as follows.

Theorem 3.3.2 ([8]) *If T is a tree of order $n \geq 4$, then $\gamma(T) - \gamma^-(T) \leq \frac{1}{5}(n - 4)$.*

Although there are classes of graphs with minus domination numbers which are
positive or zero, we remark that there are also classes of graphs with arbitrarily large
negative minus domination numbers. However, as observed in [8], if the maximum
degree of a graph is at most 5, then the minus domination number of the graph is
always zero or positive. Further, if the graph is a subcubic graph (with maximum
degree at most 3), then the minus domination number is positive. One class of so-
called "positive graphs" whose minus domination number is positive is the class of
regular graphs.

Theorem 3.3.3 ([8]) *If G is a k-regular graph of order n for some $k \geq 0$, then
$\gamma^-(G) \geq \frac{n}{k+1}$.*

Stephen Hedetniemi and his coauthors in [5] presented a variety of algorithmic
results on the complexity of minus domination in graphs, including a linear algorithm
for finding a minimum minus dominating function in a tree. Further, they showed
that the decision problems corresponding to the problem of computing $\gamma^-(G)$ and
$\Gamma^-(G)$ are both NP-complete, even when restricted to bipartite or chordal graphs.

3.4 Signed Dominating Functions

In the previous section, we discussed an application of minus domination in graphs
to networks of people in which global decisions must be made taking into account the
response or preference of individuals, which may be positive, negative, or neutral.
However often in voting, individuals are required to make either a positive or a
negative response (and do not have the option of a neutral response). To study such
networks, Stephen Hedetniemi introduced the concept of signed domination which
was first studied in [6]. For a graph $G = (V, E)$, a function $f : V \rightarrow \{-1, 1\}$ is a
signed dominating function of G if $f[v] \geq 1$ for each $v \in V$. The *signed domination
number*, denoted $\gamma_s(G)$, and the *upper signed domination number*, denoted $\Gamma_s(G)$,
of G are defined by

$$\gamma_s(G) = \min \{w(f) \mid f \text{ is a minimal signed dominating function for } G\}$$
$$\Gamma_s(G) = \max \{w(f) \mid f \text{ is a minimal signed dominating function for } G\}.$$

Fig. 3.3 The Hajós graph H

(a)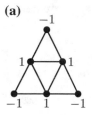

(b)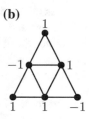

For example, the function f_1 that assigns the weights to the vertices of the Hajós graph H as illustrated in Fig. 3.3a is a minimal signed dominating function of H of weight $w(f_1) = 0$. Thus, $\gamma_s(H) \leq 0$. In fact, $\gamma_s(H) = 0$. Moreover, the function f_2 that assigns the weights to the vertices of the Hajós graph H as illustrated in Fig. 3.3b is a minimal signed dominating function of H of weight $w(f_2) = 2$. Thus, $\Gamma^-(H) \geq 2$. In fact, $\Gamma^-(H) = 2$.

The Hajós graph H in Fig. 3.3 has $\gamma_s(H) = 0$ and $\gamma(H) = 2$. On the other hand, $\gamma_s(K_{1,n}) = n$ and $\gamma(K_{1,n}) = 1$. Thus, γ and γ_s are not comparable. A lower bound on the signed domination number of a tree is established in [6].

Theorem 3.4.1 ([6]) *If T is a tree of order $n \geq 2$, then $\gamma_s(T) > \frac{1}{3}(n + 4)$, with equality if and only if T is a path on $3k + 2$ vertices for some integer $k \geq 0$.*

An analogous result to Theorem 3.3.3 also holds for the signed domination number.

Theorem 3.4.2 *If G is a k-regular graph for some $k \geq 0$, then $\gamma_s(G) \geq \frac{n}{k+1}$.*

Establishing upper bounds for the signed domination number of a regular graph proved more challenging. Zelinka [22] showed that if G is a cubic graph of order n, then $\gamma_s(G) \leq \frac{4}{5}n$. This result was generalized in [19] to all regular graphs.

Theorem 3.4.3 ([19]) *For $k \geq 2$, if G is a k-regular graph of order n, then*

$$\gamma_s(G) \leq \begin{cases} \left(\frac{(k+1)^2}{k^2+4k-1}\right)n & \text{for } k \text{ odd} \\ \\ \left(\frac{k+1}{k+3}\right)n & \text{for } k \text{ even.} \end{cases}$$

Favaron [11] showed that the bounds in Theorem 3.4.3 are sharp. Further, she improved Zelinka's result by establishing a connection between the packing number of a cubic graph and its signed domination number. A set S of vertices in G is a *packing* (also called a 2-packing in the literature) if the closed neighborhoods of vertices in S are pairwise disjoint. Equivalently, S is a packing in G if the vertices in S are pairwise at distance at least 3 apart in G. The *packing number* of G, denoted $\rho(G)$, is the maximum cardinality of a packing in G. Favaron [11] observed that if S is a packing in a cubic graph G, then the function that assigns the value -1 to the vertices of S and the value 1 to the remaining vertices is a signed dominating function of G. Conversely, if $f : V \to \{-1, 1\}$ is a signed dominating function of a graph G, then since the neighbors of every vertex of weight -1 must all have weight 1, the set of vertices of weight -1 under f form a packing in G. Therefore, $\gamma_s(G) = n - 2\rho(G)$. Favaron [11] proved the following result on the packing number of a cubic graph.

Theorem 3.4.4 ([11]) *If G is a connected cubic graph of order n different from the Petersen graph, then $\rho(G) \geq \frac{1}{8}n$.*

As an immediate consequence of Theorem 3.4.4 and the relation $\gamma_s(G) = n - 2\rho(G)$, we have the following upper bound on the signed domination number of a cubic graph.

Theorem 3.4.5 ([11]) *If G is a connected cubic graph of order n different from the Petersen graph, then $\gamma_s(G) \leq \frac{3}{4}n$.*

A variety of algorithmic results on the complexity of signed domination in graphs are presented in [14]. In particular, the authors in [14] show that the decision problems corresponding to the problem of computing $\gamma_s(G)$ and $\Gamma_s(G)$ are both NP-complete, even when restricted to bipartite or chordal graphs. A linear algorithm for finding a minimum signed dominating function in a tree is given in [13].

3.5 Real and Integer Dominating Functions

Bange, Barkauskas, Host, and Slater [1] generalized Hedetniemi's concepts of fractional, minus, and signed domination as follows. For an arbitrary subset $\mathcal{P}$ of the reals **R**, a function $f : V \to \mathcal{P}$ is a $\mathcal{P}$-*dominating function* of G if $f[v] \geq 1$ for each $v \in V$. The $\mathcal{P}$-*domination number*, denoted $\gamma_{\mathcal{P}}(G)$, is the infimum of $\mathrm{w}(f)$ taken over all $\mathcal{P}$-dominating functions f of G. Of course, this might be $-\infty$.

For example, let $\mathcal{P}$ be a subset of the reals **R**. For a real number $k \geq 1$, the function f that assigns the weights to the vertices of the Hajós graph H as illustrated in Fig. 3.4 is a minimal **R**-dominating function of H of weight $\mathrm{w}(f) = 3(1 - k)/2$. As k gets arbitrarily large, the weight of f becomes arbitrarily small, implying that $\gamma_{\mathbf{R}}(H) = -\infty$.

It turns out there is a very simple solution to determine the **R**-domination number of a graph. Let $\mathcal{P}$ be a subset of the reals **R**. For a graph $G = (V, E)$, a function

Fig. 3.4 The Hajós graph H

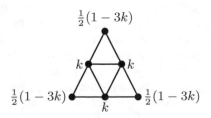

$f : V \to \mathcal{P}$ may be thought of as a vector $\vec{f}$ in $\mathcal{P}^n$. We say that $\vec{f}$ is a $\mathcal{P}$-dominating vector if and only if $N\vec{f} \geq \vec{1}$, where $\vec{1}$ denotes the all 1's vector in $\mathbf{R}^n$ and N denotes the closed neighborhood matrix of the graph G. The function $f : V \to \mathcal{P}$ is an *efficient $\mathcal{P}$-dominating function* if for every vertex v it holds that $f[v] = 1$. Equivalently, $N\vec{f} = \vec{1}$. For example, the function f that assigns the weights to the vertices of the Hajós graph H as illustrated in Fig. 3.3a is an efficient $\{-1, 1\}$-dominating function.

As first shown by Bange et al. [1], all efficient $\mathcal{P}$-dominating functions for a graph have the same weight. A simple proof of this result was given in [12].

Theorem 3.5.1 *If $\mathcal{P}$ be a subset of the reals $\mathbf{R}$ and if f_1 and f_2 are two arbitrary efficient $\mathcal{P}$-dominating functions for a graph G, then $w(f_1) = w(f_2)$.*

Proof Let N be the closed neighborhood matrix of G. Since N is symmetric,

$$\vec{f_1^{\,t}} N \, \vec{f_2} = (\vec{f_1^{\,t}} N)\, \vec{f_2} = (N\, \vec{f_1})^t \, \vec{f_2} = \vec{1}^t \, \vec{f_2} = w(f_2),$$

and

$$\vec{f_1^{\,t}} N \, \vec{f_2} = \vec{f_1^{\,t}} (N\, \vec{f_2}) = \vec{f_1^{\,t}} \, \vec{1} = w(f_1).$$

Thus, $w(f_1) - w(f_2)$. □

A function is *nonnegative* if all the function values are nonnegative. A function which is both nonnegative and efficient $\mathcal{P}$-dominating is called an NEPD-*function* (standing for **N**onnegative **E**fficient $\mathcal{P}$-**D**ominating function). For example, if G is a regular graph of degree k, then the function f that assigns to each vertex the value $1/(k + 1)$ is a NEPD-function for G noting that $f(v) > 0$ and $f[v] = 1$ for every vertex v in the graph G. In particular, a NERD-function (standing for **N**onnegative **E**fficient **R**eal **D**ominating function) in G is a nonnegative efficient real dominating function in G. The following result shows that the property of possessing a NERD-function is the key to the real domination number of a graph.

Theorem 3.5.2 ([12]) *If G is a graph, then*

$$\gamma_{\mathbf{R}}(G) = \begin{cases} w(f) & \text{if } G \text{ has a NERD-}function \text{ } f \\ -\infty & otherwise. \end{cases}$$

Proof The concept of real domination can be formulated in terms of solving the following linear programming problem:

Real Domination $\gamma_{\mathbf{R}}(G)$	**Dual**
$\min \vec{1}^t \vec{x} \left(= \min \sum_{i=1}^{n} x_i\right)$	$\max \vec{1}^t \vec{y} \left(= \max \sum_{i=1}^{n} y_i\right)$
subject to: $\begin{cases} N \cdot \vec{x} \geq \vec{1} \\ x_i \text{ unrestricted} \end{cases}$	subject to: $\begin{cases} N \cdot \vec{y} = \vec{1} \\ y_i \geq 0 \end{cases}$

The dual of the above linear programming problem is shown above on the right hand side. By taking the characteristic function of any dominating set, we obtain a feasible solution of the min problem. Hence, by linear programming duality, there are only two possible categories into which solutions to the max and min problems can fall:

(a) Both problems have feasible solutions, in which case both objective functions have the same solutions.
(b) The max problem has no feasible solution, in which case the objective function for the min problem is unbounded below.

If (a) holds, then the max problem has a feasible solution. However, every feasible solution to the max problem corresponds to a NERD-function for the graph G, and therefore the solution to the max problem in this case is a NERD-function of maximum weight. However, by Theorem 3.5.1, all NERD-functions have the same weight, implying that the solution to the max problem (and therefore the min problem) is $w(f)$, where f is an arbitrary NERD-function for G. Thus in this case, $\gamma_{\mathbf{R}}(G) = w(f)$. On the other hand, if (b) holds, then $\gamma_{\mathbf{R}}(G) = -\infty$.   □

If $\mathcal{P}$ is a subset of the reals $\mathbf{R}$ and f is a NEPD-function of a graph G, then we note that $w(f) \geq \gamma_{\mathcal{P}}(G) \geq \gamma_{\mathbf{R}}(G) = w(f)$. Hence, as an immediate corollary of Theorem 3.5.2, we have the following result.

Corollary 3.5.3 ([12]) *For any subset $\mathcal{P}$ of $\mathbf{R}$, if a graph G has a NEPD-function f, then $\gamma_{\mathcal{P}}(G) = \gamma_{\mathbf{R}}(G) = w(f)$.*

As a further consequence of Theorem 3.5.2, we have the following result.

Corollary 3.5.4 ([12]) *For any subset $\mathcal{P}$ of $\mathbf{R}$, if there exists a $\mathcal{P}$-dominating function of a graph G with total weight less than 1, then $\gamma_{\mathbf{R}}(G) = -\infty$.*

For example, we observed earlier that there is a $\{-1, 1\}$-dominating function of total weight 0 for the Hajós graph H as illustrated in Fig. 3.3a. Thus, by Corollary 3.5.4, $\gamma_{\mathbf{R}}(H) = -\infty$.

We remark that by linear algebra, if a graph has a NERD-function then it has a NEQD-function (standing for **N**onnegative **E**fficient **Q**-**D**ominating function). Thus, by Theorem 3.5.2, if G is a graph, then $\gamma_{\mathbf{Q}}(G) = w(f)$ if G has a NEQD-function f and $\gamma_{\mathbf{Q}}(G) = -\infty$, otherwise.

Integer domination, when $\mathcal{P} = \mathbf{Z}$, is also well studied. We note that if one takes a $\mathbf{Q}$-dominating function f and multiplies all the weights by the least common multiple of the weights' dominators, one obtains a $\mathbf{Z}$-dominating function. This implies that if there is a $\mathbf{Q}$-dominating function of arbitrarily negative weight, then there is a such a $\mathbf{Z}$-dominating function too. Hence, if $\gamma_{\mathbf{Q}}(G) = -\infty$, then $\gamma_{\mathbf{Z}}(G) = -\infty$.

If $\mathcal{P} = \mathbf{R}$ or $\mathcal{P} = \mathbf{Q}$, then the determination of $\gamma_{\mathcal{P}}(G)$ can be formulated in terms of solving a linear programming problem, and so can be computed in polynomial-time (see [20, 21]). However, it remains an open problem to determine the complexity of $\mathbf{Z}$-domination.

References

1. D.W. Bange, A.E. Barkauskas, L.H. Host, P.J. Slater, Generalized domination and efficient domination in graphs. Discret. Math. **159**, 1–11 (1996)
2. C. Berge, *Theory of Graphs and its Applications* (Methuen, London, 1962)
3. E.J. Cockayne, G. Fricke, S.T. Hedetniemi, C.M. Mynhardt, Properties of minimal dominating functions of graphs. Ars Combin. **41**, 107–115 (1995)
4. G.S. Domke, G.H. Fricke, R.R. Laskar, A. Majumdar, Fractional domination and related parameters. *Domination in Graphs*. Monogr. Textbooks and Applied Mathematic, vol. 209 (Dekker, New York, 1998), pp. 61–89
5. J. Dunbar, W. Goddard, S.T. Hedetniemi, M.A. Henning, A. McRae, On the algorithmic complexity of minus domination in graphs. Discret. Appl. Math. **68**, 73–84 (1996)
6. J.E. Dunbar, S.T. Hedetniemi, M.A. Henning, P.J. Slater, Signed domination in graphs. *Graph Theory, Combinatorics, and Applications* (Wiley, 1995), pp. 311–322
7. J. Dunbar, S.T. Hedetniemi, M.A. Henning, A. McRae, Minus domination in regular graphs. Discret. Math. **149**, 311–312 (1996)
8. J. Dunbar, S.T. Hedetniemi, M.A. Henning, A. McRae, Minus domination in graphs. Discret. Math. **199**, 35–47 (1999)
9. J. Dunbar, S.T. Hedetniemi, M.A. Henning, P.J. Slater, Signed domination in graphs. *Graph Theory, Combinatorics, and Applications* (Wiley, 1995), pp. 311–322
10. M. Farber, Domination, independent domination, and duality in strongly chordal graphs. Discret. Appl. Math. **7**, 115–130 (1984)
11. O. Favaron, Note: signed domination in regular graphs. Discret. Math. **158**, 287–293 (1996)
12. W. Goddard, M.A. Henning, Real and integer domination in graphs. Discret. Math. **199**, 61–75 (1999)
13. J.H. Hattingh, M.A. Henning, P.J. Slater, Three-valued k-neighbourhood domination in graphs. Australas. J. Combin. **9**, 233–242 (1994)
14. J.H. Hattingh, M.A. Henning, P.J. Slater, The algorithmic complexity of signed domination in graphs. Australas. J. Combin. **12**, 101–112 (1995)
15. T.W. Haynes, S.T. Hedetniemi, P.J. Slater, *Fundamentals of Domination in Graphs* (Marcel Dekker Inc., New York, 1998)
16. T.W. Haynes, S.T. Hedetniemi, P.J. Slater, *Domination in Graphs: Advanced Topics* (Marcel Dekker Inc., New York, 1998)
17. S.M. Hedetniemi, S.T. Hedetniemi, T.V. Wimer, Linear time resource allocation for trees. Technical report URI-014, Department Mathematical Sciences, Clemson University, 1987. Presented at southeastern conference on combinatorics, graph theory and comuting (Boca Raton, FL, 1987)
18. M.A. Henning, Dominating functions in graphs. *Domination in Graphs*. Monogr. Textbooks Pure and Applied Mathematics, vol. 209 (Dekker, New York, 1998), pp. 31–60
19. M.A. Henning, Domination in regular graphs. Ars Combin. **43**, 263–271 (1996)
20. N. Karmarkar, A new polynomial-time algorithm for linear programming. Combinatorica **4**, 373–395 (1984)
21. L.G. Khachian, A polynomial algorithm in linear programming. Dokl. Akad. Nauk SSSR **244**, 1093–1096 (1979)
22. B. Zelinka, Some remarks on domination in cubic graphs. Discret. Appl. Math. **158**, 249–255 (1996)

Chapter 4
Domination Related Parameters and Applications

In this chapter, we explore two graph theoretical concepts introduced by Stephen Hedetniemi as models for real-life applications. The first, Roman domination, is based on a historical account of a defense strategy used by the Roman Empire; and the second, alliances in graphs, models an agreement between two or more parties to work together for the common good.

4.1 Introduction

Graph theory, like many fields of mathematics, can provide precise ways of modeling real-world problems and applications. Perhaps one of Hedetniemi's strongest talents is his uncanny ability to notice a real-life application and express it as a graph theoretical concept. In many cases, his knack for asking the right questions and translating problems and applications to graph models has resulted in new research areas in graph theory. In this chapter, we briefly discuss two such ideas, namely, Roman domination in Sect. 4.2 and alliances in Sect. 4.3.

In both sections, we will use the following terminology. Let G be a graph with vertex set $V = V(G)$ and edge set $E = E(G)$ having *order* $n = |V|$. The *open neighborhood* of a vertex $v \in V$ is the set $N(v) = \{u \mid uv \in E\}$, and its *closed neighborhood* is $N[v] = N(v) \cup \{v\}$. Vertices $u \in N(v)$ are called the *neighbors* of v. A *dominating set* in a graph G is a set S of vertices of G such that every vertex in $V \setminus S$ has a neighbor in S. The *domination number* $\gamma(G)$ of a graph G is the minimum cardinality of a dominating set in G.

G. Chartrand et al., *From Domination to Coloring*, SpringerBriefs in Mathematics, https://doi.org/10.1007/978-3-030-31110-0_4

4.2 Roman Domination

Stephen Hedetniemi, along with Ernie Cockayne, Paul Dreyer, and Sandra Hedetniemi, introduced Roman domination as a graph theoretical concept in 2004. Since then over 100 papers have been published on various aspects of Roman domination in graphs. Their original paper [9] was motivated by the articles of ReVelle and Rosing [25] and Stewart [26] outlining historical strategies used to defend the Roman Empire during the reign of Emperor Constantine the Great, 274–337 AD.

At the core of the Roman Empire army were its legions, made up of highly trained and disciplined solders. These legions were stationed at various locations in the empire to defend Rome from raids and attacks by neighboring countries. A location is *protected* by a legion stationed there. Early in the fourth century Emperor Constantine changed the organizational structure and created large mobile field armies. His strategy was to economize by minimizing the number of legions at each location while still protecting all the cities. A location having no legion can be protected by a legion sent from a neighboring location. However, this presents the problem of leaving a location unprotected (without a legion) when its legion is dispatched to a neighboring location. Constantine decreed that for all cities in the Roman Empire, at most two legions should be stationed. Further, if a location having no legions is attacked, then it must be within the vicinity of at least one city at which two legions were stationed, so that one of the mobile legions can be sent to defend the attacked city. After reading this piece of the Roman Empire's history, Hedetniemi suggested the following graph theoretical model of Roman domination.

A function $f : V \to \{0, 1, 2\}$ is a *Roman dominating function*, abbreviated RD-function, on G if every vertex $u \in V$ for which $f(u) = 0$ is adjacent to at least one vertex v for which $f(v) = 2$. The *weight* of an RD-function is the value $f(V) = \sum_{u \in V} f(u)$. The *Roman domination number* $\gamma_R(G)$ is the minimum weight of an RD-function on G, and an RD-function with weight $\gamma_R(G)$ is called a γ_R-*function* of G. We represent a function $f : V \to \{0, 1, 2\}$ by the ordered partition (V_0, V_1, V_2) of V, where $V_i = \{v \in V \mid f(v) = i\}$ for $i \in \{0, 1, 2\}$. One may view Roman domination as a graph labeling problem in which each vertex labeled 0 must be adjacent to at least one vertex labeled 2.

For example, any nontrivial graph G of order n and maximum degree equal to $n - 1$ has $\gamma_R(G) = 2$ since assigning a 2 to a vertex of maximum degree and 0 to every other vertex is a γ_R-function of G. The Roman domination number of paths and cycles is given in [9].

Proposition 4.2.1 ([9]) *For paths P_n and cycles C_n with $n \geq 3$, $\gamma_R(P_n) = \gamma_R(C_n) = \lceil \frac{2n}{3} \rceil$.*

By Proposition 4.2.1, $\gamma_R(P_5) = 4$. To further illustrate Roman domination, all possible γ_R-functions (within symmetry) of the path P_5 are shown in Fig. 4.1.

Assigning 1 to each vertex of a graph G gives an RD-function of weight equal to the order n of G, so it follows that $\gamma_R(G) \leq n$ for any graph G. Chambers, Kinnersley, Prince, and West [6] gave a better bound in terms of order for connected

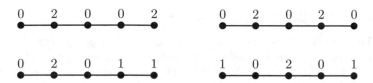

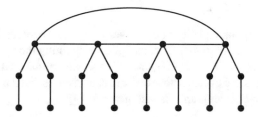

Fig. 4.1 γ_R-functions of the path P_5

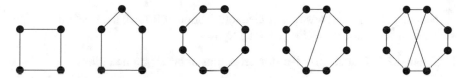

Fig. 4.2 A graph in the family $\mathcal{F}$

Fig. 4.3 The five graphs in the family $\mathcal{H}$

graphs having order at least 3, and they characterized the graphs attaining this bound. To present their result, we first define a family of graphs $\mathcal{F}$. Let F be an arbitrary connected graph, and let G be the graph of order $n = 5|V(F)|$ obtained from F by identifying each vertex of F with the center of a path P_5, where the $|V(F)|$ paths are disjoint. Let $\mathcal{F}$ be the family of all such graphs G. The graph G, where F is a cycle C_4, is illustrated in Fig. 4.2.

Theorem 4.2.2 ([6]) *If G is a connected graph of order $n \geq 3$, then $\gamma_R(G) \leq \frac{4n}{5}$, with equality if and only if G is the cycle C_5 or $G \in \mathcal{F}$.*

For graphs with minimum degree at least 2, the bound was improved in [6]. Let $\mathcal{H}$ be the family of five graphs shown in Fig. 4.3.

Theorem 4.2.3 ([6]) *If G is a connected graph of order n and minimum degree $\delta(G) \geq 2$ and $G \notin \mathcal{H}$, then $\gamma_R(G) \leq \frac{8n}{11}$.*

Liu and Chang [22] improved the bound even further for graphs having minimum degree at least 3.

Theorem 4.2.4 ([22]) *If G is a graph of order n and $\delta(G) \geq 3$, then $\gamma_R(G) \leq \frac{2n}{3}$.*

In the introductory paper on Roman domination, the following relationships with ordinary domination are observed.

Theorem 4.2.5 ([9]) *For every graph G, $\gamma(G) \leq \gamma_R(G) \leq 2\gamma(G)$.*

Proof Let $f = (V_0, V_1, V_2)$ be a γ_R-function of G. Since every vertex in V_0 has a neighbor in V_2 and the vertices of $V_1 \cup V_2$ dominate themselves, $V_1 \cup V_2$ is a dominating set of G. Thus, $\gamma(G) \leq |V_1 \cup V_2| = |V_1| + |V_2| \leq |V_1| + 2|V_2| = \gamma_R(G)$.

For the upper bound, let S be a minimum dominating set of G. The function assigning 2 to each vertex in S and 0 to every other vertex is an RD-function of G, so $\gamma_R(G) \leq 2|S| = 2\gamma(G)$. □

It is shown in [9] that equality in the lower bound of Theorem 4.2.5 is reached if and only if G is an empty graph. Extremal graphs attaining the upper bound, that is, the graphs G having $\gamma_R(G) = 2\gamma(G)$, are called *Roman graphs*. Examples of Roman graphs include paths and cycles whose order is congruent to 0 or 2 modulo 3. A simple characterization of Roman graphs is also given in [9].

Theorem 4.2.6 ([9]) *A graph G is Roman if and only if it has a γ_R-function $f = (V_0, V_1, V_2)$ with $|V_1| = 0$.*

Henning [19] gave a constructive characterization of Roman trees in 2002, but no such characterization exists for general graphs.

Problem 4.2.7 Characterize the Roman graphs of other graph families.

Favaron, Karamic, Khoeilar, and Sheikholeslami [13] determined an upper bound on the Roman domination number of a graph in terms of its order and domination number and characterized the graphs attaining this upper bound. Let F be an arbitrary nontrivial connected graph, and let G be the graph of order $n = 4|V(F)|$ obtained from F by identifying each vertex of F with an internal vertex of a path P_4 where the $|V(F)|$ paths are vertex-disjoint. Let $\mathcal{G}$ be the family of all such graphs G. For $F = C_4$, the graph G is illustrated in Fig. 4.4. The *corona* $G \circ K_1$ of G is the graph formed from G by adding for each $v \in V$, a new vertex v' and edge vv'.

Theorem 4.2.8 ([13]) *If G is a connected graph of order $n \geq 3$, then $\gamma_R(G) \leq n - \frac{\gamma(G)}{2}$, with equality if and only if G is the cycle C_4, the cycle C_5, the corona $C_4 \circ K_1$, or $G \in \mathcal{G}$.*

Interestingly, there is a relationship between the Roman domination number and the differential of a graph, a seemingly unrelated parameter. Differentials were introduced by Hedetniemi and his coauthors in [23]. For a vertex set S, the boundary of S,

Fig. 4.4 A graph in the family $\mathcal{G}$

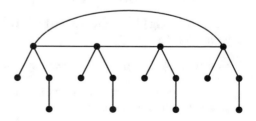

denoted $B(S)$, is the set of vertices in $V \setminus S$ that have a neighbor in S. The *differential* of the set S, denoted $\partial(S)$, is defined as $\partial(S) = |B(S)| - |S|$, and the maximum value of $\partial(S)$ for any subset S of V is the *differential* of a graph G, denoted $\partial(G)$.

Although Roman domination was introduced in 2004 and differentials in 2006, no link was known between them until 2014 when Bermudo, Fernau, and Sigarreta [4] proved that these two parameters are complementary with respect to the order of the graph as follows.

Theorem 4.2.9 ([4]) *For any graph G of order n, $\gamma_R(G) + \partial(G) = n$.*

Using this association between Roman domination and differentials in graphs, Bermudo [3] was able to improve the bound of Theorem 4.2.8 for graphs having minimum degree at least 2.

Theorem 4.2.10 ([3]) *If G is a connected graph with minimum degree $\delta(G) \geq 2$ and order $n \geq 9$, then $\gamma_R(G) \leq n - \frac{3\gamma(G)}{4}$.*

Many variations on the Roman domination number have been defined and studied. As a sample of the variants, we give a brief description of three of them in the remainder of this section.

Considering that Rome's defense strategies eventually failed, Hedetniemi and Henning [18] suggested that the Roman Empire needed a more efficient, stronger, yet leaner, defense strategy. They proposed weak Roman domination, which provided a reasonable level of defense at a cheaper cost. A vertex v with $f(v) = 0$ is said to be *undefended* with respect to f if it is not adjacent to a vertex w with $f(w) > 0$. A function f is a *weak Roman dominating function* if each vertex v with $f(v) = 0$ is adjacent to a vertex w with $f(w) > 0$, such that the function $f' = (V_0', V_1', V_2')$ defined by $f'(v) = 1$, $f'(w) = f(w) - 1$, and $f'(u) = f(u)$ for all $u \in V \setminus \{v, w\}$, has no undefended vertex. The *weak Roman domination number* $\gamma_r(G)$ equals the minimum weight of a weak Roman dominating function of G.

The weak Roman domination number of paths and cycles is given in [18].

Proposition 4.2.11 ([18]) *For paths P_n and cycles C_n with $n \geq 4$,*

$$\gamma_r(P_n) = \gamma_r(C_n) = \left\lceil \frac{3n}{7} \right\rceil.$$

Hedetniemi, along with Chellali, Haynes, and McRae, introduced a different leaner strategy called Roman $\{2\}$-domination in [7]. Roman $\{2\}$-domination was renamed *Italian domination* and studied further in [20]. A *Roman $\{2\}$-dominating function* $f : V \to \{0, 1, 2\}$ has the property that for every vertex $v \in V$ with $f(v) = 0$, $f(N(v)) \geq 2$, that is, either there is a vertex $u \in N(v)$ with $f(u) = 2$, or at least two vertices $x, y \in N(v)$ with $f(x) = f(y) = 1$. Viewed as a graph labeling problem, each vertex labeled 0 must have the labels of the vertices in its open neighborhood sum to at least 2. In terms of the Roman Empire, this defense strategy requires that every location with no legion has a neighboring location with two legions, or at least two neighboring locations with one legion each. The minimum weight of a Roman {2}-dominating function f is the *Roman {2}-domination number*, denoted $\gamma_{\{R2\}}(G)$. Consider, for example, the path P_5. In a minimum weight Roman {2}-dominating function, we can assign $1, 0, 1, 0, 1$ to the vertices for a weight of 3. But, as we have seen, the Roman domination number of P_5 is 4.

The Roman {2}-domination number of paths and cycles is given in [7].

Proposition 4.2.12 ([7]) *For paths P_n with $n \geq 1$ and cycles C_n with $n \geq 3$,*

$$\gamma_{\{R2\}}(P_n) = \left\lceil \frac{n+1}{2} \right\rceil \text{ and } \gamma_{\{R2\}}(C_n) = \left\lceil \frac{n}{2} \right\rceil.$$

The domination, weak Roman domination, Roman {2}-domination, and Roman domination numbers are related as follows.

Theorem 4.2.13 ([7]) *For every graph G,*

$$\gamma(G) \leq \gamma_r(G) \leq \gamma_{\{R2\}}(G) \leq \gamma_R(G) \leq 2\gamma(G).$$

Note that for the path $P_{10}, \gamma(P_{10}) = 4, \gamma_r(P_{10}) = 5, \gamma_{\{R2\}}(P_{10}) = 6, \gamma_R(P_{10}) = 7$, and $2\gamma(P_{10}) = 8$, implying that strict inequality for the chain of Theorem 4.2.13 is possible, even for paths. This gives rise to the following open problem.

Problem 4.2.14 Characterize the graphs G (or trees) that achieve equality in each of the inequalities of Theorem 4.2.13.

Weak Roman domination and Roman {2}-domination provide less restrictive, but somewhat weaker defense strategies than that of Roman domination. We conclude this section with a stronger approach introduced by Hedetniemi, along with Beeler and Haynes, in [2]. For the variations of Roman domination that we have seen thus far, one legion is required to defend any attacked vertex and at most two legions are stationed at any one location. Double Roman domination offers double protection, namely, that any attacked vertex can be defended by at least two legions, by allowing up to three legions to be assigned to each location. It is noted in [2] that this slight increase in the number of legions permitted per location provides the extra defense at less than the anticipated additional cost.

A function $f : V \rightarrow \{0, 1, 2, 3\}$ is a *double Roman dominating function* on a graph G if the following conditions are met:

(i) If $f(v) = 0$, then vertex v must have at least two neighbors in V_2 or one neighbor in V_3.
(ii) If $f(v) = 1$, then vertex v must have at least one neighbor in $V_2 \cup V_3$.

The *double Roman domination number* $\gamma_{dR}(G)$ equals the minimum weight of a double Roman dominating function on G.

For example, consider the complete bipartite graph $K_{2,n-2}$, for $n \geq 4$, where the Roman domination number is 3 (in a partite set of size two, assign the value 2 to one of the vertices and 1 to the other, and assign 0 to all other vertices). Increasing the value of the vertex assigned 1 to 2 gives a minimum double Roman dominating function. That is, the Roman domination number equals 3, while the double Roman domination number is only 4. Hence the defenses are doubled with an increase in expenditure of only 33%.

The double Roman domination numbers of paths and cycles are given in [1].

Proposition 4.2.15 ([1])

1. *For paths P_n,*

$$\gamma_{dR}(P_n) = \begin{cases} n & \text{if } n \equiv 0 \ (\text{mod } 3) \\ n+1 & \text{otherwise.} \end{cases}$$

2. *For cycles C_n with $n \geq 3$,*

$$\gamma_{dR}(C_n) = \begin{cases} n & \text{if } n \equiv 0, 2, 3, 4 \ (\text{mod } 6) \\ n+1 & \text{otherwise.} \end{cases}$$

Double Roman domination is related to domination and Roman domination as follows.

Theorem 4.2.16 ([2]) *For any nontrivial connected graph G,*

1. $\gamma_R(G) < \gamma_{dR}(G) < 2\gamma_R(G)$, *and*
2. $\gamma(G) \leq \gamma_R(G) \leq 2\gamma(G) \leq \gamma_{dR}(G) \leq 3\gamma(G)$.

4.3 Alliances

An alliance is generally thought of as a pact or formal treaty between two or more parties, made in order to unite for a common cause. Stephen Hedetniemi, along with Sandra Hedetniemi and Petter Kristiansen, introduced several types of alliances in graphs to model agreements between nations for mutual support (see [17]). For example, they considered defensive alliances during times of war, where the allies agree to join forces if one or more of them are attacked, and also offensive alliances in times of peace, where allies join forces in order to keep peace. In addition to national defense coalitions, applications of alliances are widespread in nature from social and business associations to political and scientific groupings. As with Roman domination, the study of alliances in graphs has become a popular area of research with around 100 papers published since its inception in 2004. In this section, we present a few different alliances, focusing on ones suggested by Stephen Hedetniemi. For more information on alliances, we refer the reader to the three survey papers [24, 27, 28].

Recall that for a vertex set S, the boundary of S, denoted $B(S)$, is the set of vertices in $V \setminus S$ that have a neighbor in S, and the differential of S is defined as $\partial(S) = |B(S)| - |S|$.

Defensive and offensive alliances were first introduced in [17]. A nonempty set of vertices $S \subseteq V$ is called a *defensive alliance* if for every $v \in S$, $|N[v] \cap S| \geq |N(v) \cap (V \setminus S)|$. The minimum cardinality of a defensive alliance of G is denoted by $a(G)$. In terms of application of a defensive alliance S, it is reasonable to think that each vertex in S is in alliance with its neighbors in S against its neighbors in $\partial(S)$. For the set S as a whole, since an attack by the vertices of $\partial(S)$ on a defensive alliance S can result in no worse than a "tie" (assuming strength in numbers), the vertices in S can "successfully" defend against the vertices of $\partial(S)$.

A nonempty set S is an *offensive alliance* if for every vertex $v \in \partial(S)$, $|N(v) \cap S| \geq |N[v] \cap (V \setminus S)|$. The minimum cardinality of an offensive alliance of G is denoted by $a_o(G)$. In this case, the vertices in S can "successfully" attack $\partial(S)$.

The offensive alliance and defensive alliance numbers are equal for a complete graph, that is, $a(K_n) = a_o(K_n) = \lceil \frac{n}{2} \rceil$. Note that any vertex of degree at most 1 is a defensive alliance. It is shown in [17] that $a(G) = 1$ if and only if G has a vertex of degree 0 or 1, and it is shown in [12] that $a_o(G) = 1$ if and only if G is a star. The alliance numbers for paths and cycles follow.

Proposition 4.3.1 *For paths P_n and cycles C_n with $n \geq 3$,*

1. *([17]) $a(P_n) = 1$ and $a(C_n) = 2$,*
2. *([12]) $a_o(P_n) = \lfloor \frac{n}{2} \rfloor$ and $a_o(C_n) = \lceil \frac{n}{2} \rceil$.*

Paths and cycles provide examples where the offensive alliance number can be larger than the defensive alliance number. To see that these two numbers are incomparable, consider the complete bipartite graphs $K_{r,s}$, where $2 \leq r \leq s$. Then $a(K_{r,s}) = \lfloor \frac{r}{2} \rfloor + \lfloor \frac{s}{2} \rfloor$, while $a_o(K_{r,s}) = \lceil \frac{r+1}{2} \rceil$. Thus, the defensive alliance number is larger than the offensive alliance number for $K_{r,s}$ when $r \geq 4$.

Odile Favaron, Gerd Fricke, Wayne Goddard, Sandra Hedetniemi, Stephen Hedetniemi, Petter Kristiansen, Renu Laskar, and Duane Skaggs [12] determined an upper bound on the offensive alliance number of a graph in terms of its order. An edge uv of G is called *monochromatic* if u and v are assigned the same color in a given vertex coloring of G.

Theorem 4.3.2 ([12]) *If G is a graph of order $n \geq 2$, then $a_o(G) \leq \frac{2n}{3}$.*

Proof Since the result is trivial if G has an isolated vertex, we may assume that the minimum degree of G is at least 1. Color the vertices of V with three colors such that the number of monochromatic edges is minimized. Then any vertex is incident with at least double the number of non-monochromatic edges as monochromatic edges. (If a green vertex has more green neighbors than red neighbors, then we can recolor it red, a contradiction.) Thus, the union of any two color classes is an offensive alliance, implying that $a_o \leq \frac{2n}{3}$. □

As observed in [12], the bound of Theorem 4.3.2 is sharp and is attained by the triangle K_3, the complete tripartite graph $K_{2,2,2}$, and the graph formed from three disjoint triangles T_1, T_2, and T_3 by adding three edges so that there is a triangle containing one vertex from each of T_1, T_2, and T_3.

Stephen Hedetniemi and his coauthors determined a tight upper bound on the defensive alliance number of a graph in terms of its order in [14]. A *balanced bipartition* of G is a partition of V into two sets A and B, where $|A| = \lceil \frac{n}{2} \rceil$ and $B = \lfloor \frac{n}{2} \rfloor$. An edge joining a vertex in A with a vertex in B is called an *AB-edge*.

Theorem 4.3.3 ([14]) *If G is a connected graph of order $n \geq 2$, then $a(G) \leq \lceil \frac{n}{2} \rceil$.*

Proof The result is trivial if G has a vertex of degree at most 1. Among all balanced bipartitions (A, B) of V, let $\pi = (A, B)$ be one that minimizes the number of AB-edges. Then $|A| = \lceil \frac{n}{2} \rceil$ and $B = \lfloor \frac{n}{2} \rfloor$. If A or B is a defensive alliance, then the result holds. Hence, assume that neither A nor B is a defensive alliance. Thus, there exist vertices $a \in A$ and $b \in B$ such that $|N[a] \cap A| < |N(a) \cap B|$ and $|N[b] \cap B| < |N(b) \cap A|$. Let $A' = (A \setminus \{a\}) \cup \{b\}$ and $B' = (B \setminus \{b\}) \cup \{a\}$. But then $\pi' = (A', B')$ is a balanced bipartition of V with fewer $A'B'$-edges than AB-edges, contradicting our choice of π. $\qquad\square$

Brigham, Dutton, Haynes, and Hedetniemi [5] studied alliances that are both defensive and offensive, which they called powerful alliances. That is, a vertex set S is a *powerful alliance* if for every vertex $v \in N[S]$, $|N[v] \cap S| \geq |N[v] \cap (V \setminus S)|$. An alliance S of any type (defensive, offensive, or powerful) is called *global* if S is a dominating set. Much of the research on alliances has involved global alliances. See [8, 11, 15, 16, 21, 29], for example.

We conclude this section with a relation of alliances called distribution centers, which were also first suggested by Hedetniemi. In business, a distribution center for a set of products is a structure or a group of units used to store goods that are to be distributed to retailers, to wholesalers, or directly to consumers. Distribution centers are usually thought of as being demand driven.

Hedetniemi, along with Desormeaux, Haynes, and Moore, defined a distribution center in a graph to model a supply and demand situation (see [10]). Formally, a nonempty set of vertices S is a *distribution center* of G if every vertex $v \in \partial(S)$ is adjacent to a vertex $u \in S$ with $|N[u] \cap S| \geq |N[v] \cap (V \setminus S)|$. The minimum cardinality of a distribution center of a graph G is the *distribution center number* $dc(G)$, and a distribution center of G with cardinality $dc(G)$ is called a dc-set of G.

Let S be a nonempty set of vertices of G. If $u \in S$, $v \in V \setminus S$, and $v \in N(u)$, such that $|N[u] \cap S| \geq |N[v] \cap (V \setminus S)|$, then we say that u supplies the demand of v. One way to look at a distribution center S is to think of a vertex $v \in \partial(S)$ and its neighbors in $V \setminus S$ as needing some amount of resource units, one unit per vertex, while each vertex in S is able to supply one unit of the resource. Thus, a vertex in $\partial(S)$ makes a demand on the distribution center S and is supplied by one of its neighbors in S. Vertex v can ask a vertex $u \in S \cap N(v)$ to deliver $|N[v] \cap (V \setminus S)|$ units. Vertex u can provide this amount only if vertex u can receive from itself and its neighbors in S at least this number, that is, $|N[u] \cap S| \geq |N[v] \cap (V \setminus S)|$. Hence, such a set S models a distribution center that is capable of providing *two-day delivery* to any vertex (customer) in $\partial(S)$: on day 1, each neighbor of $u \in S$ ships one unit of resource to u, and then, on day 2, vertex u ships $|N[v] \cap (V \setminus S)|$ units of resource to its neighbor $v \in \partial(u)$. Thus, a distribution center is a type of an alliance between the vertices of S to service the vertices in $\partial(S)$.

For the star $K_{1,n-1}$, $\mathrm{dc}(K_{1,n-1}) = 1$ since each leaf has a demand of one and the center vertex can supply the demand. Similarly, any two adjacent vertices on a path or a cycle form a distribution center, so we have the following result.

Proposition 4.3.4 ([10])

1. For a cycle C_n with $n \geq 3$, $\mathrm{dc}(C_n) = 2$.
2. For a path P_n with $n \geq 4$, $\mathrm{dc}(P_n) = 2$.

Although distribution centers and offensive alliances are similar concepts, the corresponding parameters can easily be shown to be incomparable. To see this, note that for cycles C_n with $n \geq 5$, $a_o(C_n) = \lceil \frac{n}{2} \rceil > 2 = \mathrm{dc}(C_n)$. On the other hand, for the complete bipartite graph $K_{r,s}$ with $1 \leq r \leq s$, $a_o(K_{r,s}) = \lceil \frac{r+1}{2} \rceil$, while $\mathrm{dc}(K_{r,s}) = r$. Thus, $a_o(K_{r,s}) < \mathrm{dc}(K_{r,s})$ for $r \geq 3$.

Our next result shows that the domination number is an upper bound on the distribution center number of any tree. This bound does not necessarily hold for general graphs. For example, for the complete graph K_n, the domination number is one, while the distribution center number is $\lceil n/2 \rceil$, showing that the distribution center number can be much larger than the domination number.

For a tree T rooted at a vertex r and a vertex $v \neq r$ of T, let T_v denote the subtree of T rooted at v, consisting of v and its descendants in T. Further, let $T - T_v$ denote the tree rooted at r formed by removing the subtree T_v from T. Note that the vertices and edges of T_v and the edge from v to its parent in T are removed from T to form $T - T_v$ and that $T - T_v$ is a tree. In a rooted tree, a support vertex all of whose children are leaves is called a *terminal support vertex*. Let $\mathrm{diam}(T)$ denote the diameter of T.

Theorem 4.3.5 ([10]) If T is a tree, then $\mathrm{dc}(T) \leq \gamma(T)$.

Proof Let T be a tree of order n. We proceed by induction on n. Note that if T is the trivial graph or the star $K_{1,n-1}$, then $\mathrm{dc}(T) = 1 = \gamma(T)$, and if T is the double star $S_{p,q}$ (where $1 \leq p \leq q$), then $\mathrm{dc}(T) = 2 = \gamma(T)$. Hence, we can assume that $\mathrm{diam}(T) \geq 4$. This implies that $n \geq 5$ and $\gamma(T) \geq 2$.

Assume that any tree T' with order $n' < n$ has $\mathrm{dc}(T') \leq \gamma(T')$.

Let r and v be two vertices at $\mathrm{diam}(T)$ apart, and root T at r. Necessarily, r and v are leaves of T. Let u be the parent of v, w the parent of u, and x the parent of w. Note that by our choice of v, u is a terminal support vertex in T. If w has degree 2, then $\{u, v\}$ is a distribution center of T implying that $\mathrm{dc}(T) = 2 \leq \gamma(T)$.

Hence, we can assume that the degree of w is at least 3. By our choice of u, every child of w is either a leaf or a terminal support vertex. Let $T' = T - T_u$. Let D be a minimum dominating set of T containing the support vertices of T (which is always possible since either the support vertex or its adjacent leaves must be in every dominating set), and let D' be a restriction of D onto T'. Since u is a support vertex of T, we can assume that u is in D. If w has a leaf neighbor, then w is in D; otherwise w is dominated by a child in D' that is a support vertex. Hence, $D' = D \setminus \{u\}$ is a dominating set of T' and so $\gamma(T') \leq \gamma(T) - 1$.

Fig. 4.5 Graph G with $\mathrm{dc}(G) < \mathrm{udc}(G)$

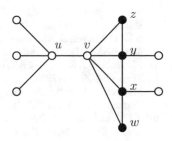

Let S' be a dc-set of T'. If $N[w] \cap S' \neq \emptyset$, then $S' \cup \{u\}$ is a distribution center of T. Thus, $\mathrm{dc}(T) \leq \mathrm{dc}(T') + 1 \leq \gamma(T') + 1 \leq \gamma(T)$, as desired. Hence, assume that $N[w] \cap S' = \emptyset$. But then S' is a distribution center of T. Therefore, $\mathrm{dc}(T) \leq |S'| = \mathrm{dc}(T') \leq \gamma(T') \leq \gamma(T) - 1 < \gamma(T)$, and the result follows. $\quad\square$

We conclude with three open problems, two of which are listed in [10].

Problem 4.3.6 Characterize the trees T having $\mathrm{dc}(T) = \gamma(T)$.

Problem 4.3.7 Investigate the maximum number of pairwise disjoint distribution centers in a graph G.

Problem 4.3.8 Let S be a nonempty set of vertices of G. An edge uv in $[S, \partial(S)]$, where $u \in S$ and $v \in \partial(S)$, is called a *supply line* if $|N[u] \cap S| \geq |N[v] \cap (V \setminus S)|$. A distribution center $S \subseteq V$ is a universal distribution center if every edge $uv \in [S, \partial(S)]$ is a supply line, that is, $|N[u] \cap S| \geq |N[v] \cap (V \setminus S)|$ for every edge $uv \in [S, \partial(S)]$. The universal distribution center number $udc(G)$ is the minimum cardinality of a universal distribution center set of G. Not all distribution centers are universal distribution centers. For example, the graph G in Fig. 4.5 has $\mathrm{dc}(G) = 3 < 4 = udc(G)$. To see this, note that the set $S = \{u, v, y\}$ is a distribution center, but not a universal distribution center since xy is not a supply line. The set $S' = \{w, x, y, z\}$ is a universal distribution center. Study universal distribution centers.

References

1. H. Abdollahzadeh Ahangar, M. Chellali, M.S. Sheikholeslami, On the double Roman domination in graphs. Discret. Appl. Math. **232**, 1–7 (2017)
2. R.A. Beeler, T.W. Haynes, S.T. Hedetniemi, Double Roman domination. Discret. Appl. Math. **211**, 23–29 (2016)
3. S. Bermudo, On the differential and Roman domination number of a graph with minimum degree two. Discret. Appl. Math. **232**, 64–72 (2017)
4. S. Bermudo, H. Fernau, J.M. Sigarreta, The differential and the Roman domination number of a graph. Appl. Anal. Discret. Math. **8**, 155–171 (2014)

5. R.C. Brigham, R.D. Dutton, T.W. Haynes, S.T. Hedetniemi, Powerful alliances in graphs. Discret. Math. **309**, 2140–2147 (2009)
6. E.W. Chambers, B. Kinnersley, N. Prince, D.B. West, Extremal problems for Roman domination. SIAM J. Discret. Math. **23**, 1575–1586 (2009)
7. M. Chellali, T.W. Haynes, S.T. Hedetniemi, A. McRae, Roman {2}-domination. Discret. Appl. Math. **204**, 22–28 (2016)
8. M. Chellali, L. Volkmann, Independence and global offensive alliance in graphs. Australas. J. Combin. **47**, 125–131 (2010)
9. E.J. Cockayne, P.M. Dreyer Sr., S.M. Hedetniemi, S.T. Hedetniemi, Roman domination in graphs. Discret. Math. **278**, 11–22 (2004)
10. W.J. Desormeaux, T.W. Haynes, S.T. Hedetniemi, Distribution centers in graphs. Discret. Appl. Math. **243**, 186–193 (2018)
11. O. Favaron, Global alliances and independent domination in some classes of graphs. Electron. J. Combin. **15**, Research Paper 123, 9 (2008)
12. O. Favaron, G.H. Fricke, W. Goddard, S.M. Hedetniemi, S.T. Hedetniemi, P. Kristiansen, R. Laskar, R.D. Skaggs, Offensive alliances in graphs. Discuss. Math. Graph Theory **24**, 263–275 (2004)
13. O. Favaron, H. Karami, R. Khoeilar, S.M. Sheikholeslami, On the Roman domination number of a graph. Discret. Math. **309**, 3447–3451 (2009)
14. G.H. Fricke, L.M. Lawson, T.W. Haynes, S.M. Hedetniemi, S.T. Hedetniemi, A note on defensive alliances in graphs. Bull. Inst. Combin. Appl. **38**, 37–41 (2003)
15. A. Harutyunyan, Global offensive alliances in graphs and random graphs. Discret. Appl. Math. **164**, 522–526 (2014)
16. T.W. Haynes, S.T. Hedetniemi, M.A. Henning, Global defensive alliances in graphs. Electron. J. Combin. **10**, Research Paper 47, 13 (2003)
17. S.M. Hedetniemi, S.T. Hedetniemi, P. Kristiansen, Alliances in graphs. J. Combin. Math. Combin. Comput. **48**, 157–177 (2004)
18. S.T. Hedetniemi, M.A. Henning, Defending the Roman Empire - a new strategy. Discret. Math. **266**, 239–251 (2003)
19. M.A. Henning, A characterization of Roman trees. Discuss. Math. Graph Theory **22**, 325–334 (2002)
20. M.A. Henning, W.F. Klostermeyer, Italian domination in trees. Discret. Appl. Math. **217**, 557–564 (2017)
21. N. Jafari Rad, A note on the global offensive alliances in graphs. Discret. Appl. Math. **250**, 373–376 (2018)
22. C. Liu, G.J. Chang, Upper bounds on Roman domination numbers of graphs. Discret. Math. **312**, 1386–1391 (2012)
23. J.L. Mashburn, T.W. Haynes, S.M. Hedetniemi, S.T. Hedetniemi, P.J. Slater, Differentials in graphs. Util. Math. **69**, 43–54 (2006)
24. K. Ouazine, H. Slimani, A. Tari, Alliances in graphs: parameters, properties and applications-a survey. AKCE Int. J. Graphs Combin. **15**, 115–154 (2018)
25. C.S. ReVelle, K.E. Rosing, Defendens imperium romanum: a classical problem in military strategy. Am. Math. Mon. **107**, 585–594 (2000)
26. I. Stewart, Defend the Roman Empire! Sci. Am. **281**, 136–139 (1999)
27. I.G. Yero, J.A. Rodriguez-Velázquez, A survey on alliances and related parameters in graphs. Electron. J. Graph Theory Appl. **2**, 70–86 (2014)
28. I.G. Yero, J.A. Rodriguez-Velázquez, A survey on alliances in graphs: defensive alliances. Util. Math. **105**, 141–172 (2017)
29. A. Yu, An inequality on global alliances for trees. Discret. Appl. Math. **185**, 227–229 (2015)

Chapter 5
Distance-Defined Subgraphs

In a connected graph G, there is a path connecting every two vertices of G; in fact, there may be several such paths. For vertices u and v of G, the length of a shortest $u - v$ path in G is the distance between u and v. For every vertex v of G, it is often of interest to know the distance from v to a vertex of G farthest from v (the eccentricity of v). The total distance of v is the sum of the distances from v to all vertices of G. The vertices of a connected graph having minimum eccentricity, those having maximum eccentricity, and those having minimum total distance and the subgraphs induced by these three sets of vertices are the primary topics of this chapter.

5.1 Distance Parameters

The *distance* $d(u, v)$ from a vertex u to a vertex v in a connected graph G is the length of a shortest $u - v$ path in G. A $u - v$ path of length $d(u, v)$ is called a $u - v$ *geodesic*. The *eccentricity* $e(v)$ of a vertex v in G is the distance from v to a vertex farthest from v, that is,

$$e(v) = \max\{d(v, x) : x \in V(G)\}.$$

The *diameter* $\operatorname{diam}(G)$ of G is the greatest eccentricity among the vertices of G, while the *radius* $\operatorname{rad}(G)$ of G is the smallest eccentricity among the vertices of G. The diameter of G is therefore the greatest distance between any two vertices of G. The distance d defined above satisfies the following properties in a connected graph G: (1) $d(u, v) \geq 0$ for every two vertices u and v of G; (2) $d(u, v) = 0$ if and only if $u = v$; (3) $d(u, v) = d(v, u)$ for all $u, v \in V(G)$ (the *symmetric property*); (4) $d(u, w) \leq d(u, v) + d(v, w)$ for all $u, v, w \in V(G)$ (the *triangle inequality*). Consequently, d is a *metric* on $V(G)$ and $(V(G), d)$ is a *metric space*. The following result gives a relationship between the radius and diameter of a nontrivial connected graph.

© The Author(s), under exclusive license to Springer Nature Switzerland AG 2019
G. Chartrand et al., *From Domination to Coloring*, SpringerBriefs in Mathematics,
https://doi.org/10.1007/978-3-030-31110-0_5

Theorem 5.1.1 *For every nontrivial connected graph G,*

$$\mathrm{rad}(G) \leq \mathrm{diam}(G) \leq 2\,\mathrm{rad}(G).$$

In 1973, Ostrand showed not only that the bounds for the diameter of a graph stated in Theorem 5.1.1 are sharp but that each pair a, b of positive integers satisfying these inequalities can be realized as the radius and diameter of some graph.

Theorem 5.1.2 ([12]) *For every two positive integers a and b with a $\leq$ b $\leq$ 2a − 2, there exist graphs of radius a and diameter b. Furthermore, the minimum order of such a graph is a + b.*

The *total distance* $\mathrm{td}(v)$ of a vertex v in a connected graph G is defined by

$$\mathrm{td}(v) = \sum_{w \in V(G)} d(v, w).$$

This concept was introduced by Harary [4] in 1959. Studying the total distances of the vertices of a connected graph G of order n is essentially the same as studying average distances of the vertices as the *average distance* from a vertex v to the vertices of G is $\mathrm{td}(v)/n$.

A vertex v of a connected graph G with $e(v) = \mathrm{rad}(G)$ is called a *central vertex* of G and the subgraph induced by the central vertices of G is the *center* $\mathrm{Cen}(G)$ of G. A vertex v in a connected graph G is called a *median vertex* of G if v has the minimum total distance among the vertices of G. Equivalently, v is a median vertex if v has the minimum average distance to the vertices of G. The *median* $\mathrm{Med}(G)$ of G is the subgraph of G induced by its median vertices. Therefore, the center and median of a connected graph G are subgraphs of G that might be considered as two possible interpretations of the "middle" of G. For example, consider the graph G of Fig. 5.1, where each vertex in the second figure is labeled with the eccentricity of the vertex, while each vertex in the third figure is labeled with the total distance of the vertex. Thus, the center of G contains the two vertices u and v and the median of G consists of the single vertex w.

The concepts of center and median have an abundance of real-life applications. For example, suppose that the goal is to find an optimal location for an emergency facility in a certain city, such as a police station, a fire station, or a medical center. Then we might want to minimize the distance from any location in the city that is farthest from this facility. If the street system of the city is modeled by a graph G, then a possible location for such an emergency facility might be somewhere in the city that corresponds to the center of G. On the other hand, if the goal is to determine a location for a service facility in the city, such as a post office, a shopping center, or a bank, then we might want to minimize the average distance from all customers in the city to the facility. An optimal location in this case might be somewhere in the city that corresponds to the median of G. For example, suppose that the graph G of Fig. 5.1 represents the street system of a community, where the edges are streets and vertices

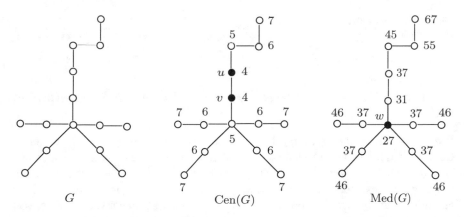

Fig. 5.1 The center and median of a graph

are street intersections. If the community wants to build an emergency facility at a location that minimizes the drive from any street intersection to this facility, then an appropriate location for this emergency facility would be somewhere along the street joining u and v. If the community wants to build a service facility at a location that minimizes the average drive from all street intersections to the facility, then the likely location for this service facility is the street intersection corresponding to w. We now consider these two distance-defined subgraphs—the center and the median.

5.2 The Center of a Graph

A fundamental question concerning centers is that of determining those graphs that can be the center of some graph. By a result initially published in a paper by Buckley, Miller, and Slater [3], Stephen Hedetniemi showed that there is no restriction of which graphs can be the center of some graph.

Theorem 5.2.1 *For every graph H, there exists a connected graph G such that*

$$\mathrm{Cen}(G) \cong H.$$

Proof Let G be the graph constructed from H by first adding two new vertices u and v to H and joining them to every vertex of H but not to each other. The construction of G is completed by adding two other vertices u_1 and v_1, where u_1 is joined to u and v_1 is joined to v (see Fig. 5.2). Since $e(u_1) = e(v_1) = 4$, $e(u) = e(v) = 3$, and $e(x) = 2$ for every vertex x in H, it follows that $V(H)$ is the set of central vertices of G and so $\mathrm{Cen}(G) = G[V(H)] = H$. ∎

A property concerning the location of the center of every connected graph was observed by Harary and Norman.

Fig. 5.2 A graph with a
given center

$$G:$$

Theorem 5.2.2 ([5]) *The center of every connected graph G lies in a single block of G.*

Proof Assume, to the contrary, that there is a connected graph G whose center Cen(G) does not lie within a single block of G. Then G has a cut-vertex v such that $G - v$ contains components G_1 and G_2, each of which contains vertices of Cen(G). Let u be a vertex such that $d(u, v) = e(v)$ and let P_1 be a $v - u$ geodesic. At least one of G_1 and G_2, say G_2, contains no vertices of P_1. Let w be a vertex of Cen(G) belonging to G_2, and let P_2 be a $w - v$ geodesic. The paths P_1 and P_2 together form a $u - w$ path P_3, which is necessarily a $u - w$ path of length $d(u, w)$. However then, $e(w) > e(v)$, which is a contradiction. Thus, Cen(G) lies in a single block of G. ∎

Among the research topics involving centers are those of determining possible centers of graphs belonging to some familiar classes.

By Theorem 5.2.2, the center of every graph lies in a block. Since the only blocks of a tree are K_2, the only possible centers are K_1 and K_2. This observation provides a proof of a classical theorem of Jordan. A tree is called *central* if its center is K_1 and *bicentral* if its center is K_2.

Theorem 5.2.3 ([8]) *Every tree is either central or bicentral.*

A graph G is *planar* if it can be embedded in the plane, that is, G can be drawn in the plane without any two of its edges crossing. Such a drawing is also called a *planar embedding* of G. A graph G is *outerplanar* if there exists a planar embedding of G so that every vertex of G lies on the boundary of the exterior region. An outerplanar graph G is *maximal outerplanar* if the addition to G of any edge joining two nonadjacent vertices of G results in a graph that is not outerplanar. Therefore, if G is a maximal outerplanar graph of order at least 3, then there is a planar embedding of G where the boundary of the exterior region contains every vertex of G and the boundary of every other region is a triangle. In 1980, Proskurowski obtained the following characterization of those graphs that can be the center of a maximal outerplanar graph.

Theorem 5.2.4 ([13]) *If G is a maximal outerplanar graph, then the center of G is isomorphic to one of the seven graphs in* Fig. 5.3.

A graph is a $C_{(n)}$-*tree*, $n \geq 3$, if it can be constructed in the following manner. Let C be an n-cycle, which we denote by G_0. The graph G_1 is constructed from G_0 by adding a new n-cycle C' to G_0 and identifying an edge of G_0 and an edge of C', which results in the graph G_1. Proceeding recursively, for an integer $k \geq 2$, the graph G_k is constructed from G_{k-1} by adding a new n-cycle C' to G_{k-1} and identifying an edge of G_{k-1} and an edge of C'. Each such graph G_k is called a

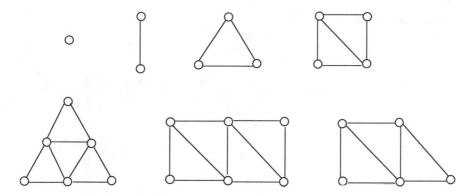

Fig. 5.3 The seven possible centers of maximal outerplanar graphs

$C_{(n)}$-tree. Proskurowski [14] determined all those graphs that can be the center of a $C_{(3)}$-tree. Mitchell and Hedetniemi determined all those graphs that can be the center of a $C_{(4)}$-tree and Mitchell, Hedetniemi, and Slater investigated centers of $C_{(n)}$-trees for all integers $n \geq 5$ (see [6]).

Theorem 5.2.1 has led to a succession of research topics involving not only the center of a graph but numerous other distance-defined subgraphs in a connected graph. A more general interpretation of center was introduced by Stephen Hedetniemi, along with Mitchell and Cockayne [10] and by Slater [16]. Let P be a path in a tree T. For a vertex v in T, the *distance between v and P* is defined as

$$d(v, P) = \min\{d(v, u) : u \in V(P)\}.$$

Therefore, if $v \in V(P)$, then $d(v, P) = 0$. The *eccentricity $e(P)$ of the path P* is defined as

$$e(P) = \max\{d(v, P) : v \in V(T)\}.$$

The *path radius* prad(T) of a tree T is defined as

$$\text{prad}(T) = \min\{e(P) : P \text{ is a path in } T\}.$$

A path P in a tree T is *minimal with respect to its eccentricity* if every proper subpath of P has greater eccentricity. A minimal path P^* of T for which $e(P^*) = \text{prad}(T)$ is called a *central path* of T. The *path center* of T is the union of all central paths of T.

These concepts are illustrated for the tree T of Fig. 5.4. Let $P^* = (u, x, y, z, v)$. While $e(P^*) = 3$, no path in T has eccentricity 2. Since the eccentricity of every proper subpath of P^* exceeds 3, it follows that P^* is minimal with respect to its eccentricity. Thus, prad$(T) = 3$ and P^* is the only central path of T. Consequently, P^* is the path center of T.

Fig. 5.4 The path center of
a tree

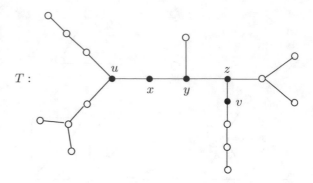

$T:$

The fact that the path center of the tree T in Fig. 5.4 consists of a single path is not surprising. Hedetniemi, along with Mitchell and Cockayne, obtained the following result.

Theorem 5.2.5 ([10]) *The path center of every tree consists of a single path.*

Proof Assume, to the contrary, that there exists a tree T whose path center does not consist of a single path. Then there are two distinct paths P_1 and P_2 in the center of T. We consider two cases, depending on whether $V(P_1)$ and $V(P_2)$ are disjoint.

   Case 1. $V(P_1) \cap V(P_2) = \emptyset$. Let P be the unique path from P_1 to P_2 in T, say P is a $u_1 - u_2$ path where $u_1 \in V(P_1)$ and $u_2 \in V(P_2)$, that is, $V(P) \cap V(P_i) = \{u_i\}$ for $i = 1, 2$. Let T_i $(i = 1, 2)$ be the component (tree) containing P_i in the forest $T - E(P)$ and let $x_i \in V(T_i)$ such that

$$d(x_i, u_i) = \max\{d(x, u_i) : x \in V(T_i)\}.$$

We show that the $x_1 - x_2$ path P^* has the property that

$$e(P^*) < e(P_1) = e(P_2).$$

If $y \in V(T_1)$, then

$$d(y, P^*) \leq d(y, u_1) \leq d(x_1, u_1) < d(x_1, P_2) \leq e(P_2).$$

Similarly, if $y \in V(T_2)$, then $d(y, P^*) < e(P_1)$. If z is a vertex of T whose path to P contains no vertex of T_1 or T_2, then

$$d(z, P^*) < d(z, P_i) \leq e(P_i) \quad \text{for } i = 1, 2.$$

If $z \in V(P)$, then $d(z, P^*) = 0$. Therefore, $e(P^*) < e(P_1) = e(P_2)$, which is a contradiction.

   Case 2. $V(P_1) \cap V(P_2) \neq \emptyset$, say $V(P_1) \cap V(P_2) = \{u_1, u_2, \ldots, u_k\}$ for some positive integer k. Since P_1 and P_2 intersect in a path, we may assume that this path is $(u_1, u_2, \ldots, u_k)$. Suppose that P_1 has an end-vertex $x_1 \notin U$. Since P_1 is minimal with respect to its eccentricity, there exists a vertex $y_1 \in V(T)$ such that the shortest path from y_1 to P_1 meets P_1 at a vertex x_1 and $d(y_1, x_1) = e(P_1)$. However, P_1 has no end-vertex that is not in U. The same is true for P_2. Thus, $P_1 = P_2$. ∎

Since graph theory has important applications in computer science, the area of graph algorithms has been one of the fastest growing areas of research in graph theory. Algorithms have been developed for various types of distance-defined subgraphs of a connected graph or a network (see [2]). In particular, Hedetniemi, along with Mitchell and Cockayne [10], and Slater [16] have developed algorithms to find path centers in trees.

5.3 The Median of a Graph

While Hedetniemi proved that every graph is the center of some graph, his former doctoral student Peter Slater [15] showed in 1980 that every connected graph is also the median of some graph. Later, Miller [9] found a simpler proof of this fact.

Theorem 5.3.1 ([9, 15]) *For every graph H, there exists a connected graph G such that*

$$\text{Med}(G) \cong H.$$

Proof Let $V(H) = \{v_1, v_2, \ldots, v_n\}$ and let G be that graph constructed from H by adding n new vertices $u_1, u_2, \ldots, u_n$ and joining u_i to v_i as well as all vertices *not* adjacent to v_i in H. For each integer i with $1 \le i \le n$, $\text{td}(v_i) = 3n - 2$ and $\text{td}(u_i) = 3n - 2 + \deg_G v_i$. Since $\deg_G v_i \ge 1$, it follows that $\text{Med}(G) \cong H$. ∎

In 1985, Truszyński established the following analogue of Theorem 5.2.2 for medians.

Theorem 5.3.2 ([17]) *The median of every connected graph G lies in a single block of G.*

A consequence of Theorem 5.3.2 is then an analogue of Theorem 5.2.3.

Corollary 5.3.3 *The median of every tree is either K_1 or K_2.*

It was seen, in Fig. 5.1, that the center and median of a connected graph are not always identical. It may be expected that these two subgraphs are located quite close to each other. Novotny and Tian showed that not only can the center and median be close to each other but can overlap in any prescribed manner.

Theorem 5.3.4 ([11]) *For every two graphs H_1 and H_2 and a graph H that is an induced subgraph of both H_1 and H_2, there exists a connected graph G such that*

$$\text{Cen}(G) \cong H_1, \text{Med}(G) \cong H_2, \text{ and } \text{Cen}(G) \cap \text{Med}(G) \cong H.$$

Holbert [7] then showed that connected graphs exist whose center and median are arbitrarily far apart. For two subgraphs F and H in a connected graph G, the *distance $d(F, H)$* between F and H is defined as

$$d(F, H) = \min\{d(u, v) : u \in V(F), v \in V(H)\}.$$

Thus, $d(F, H) = 0$ if $V(F) \cap V(H) \neq \emptyset$.

Theorem 5.3.5 ([7]) *For every two graphs H_1 and H_2 and every positive integer k, there exists a connected graph G such that*

$$\mathrm{Cen}(G) \cong H_1, \mathrm{Med}(G) \cong H_2, \text{ and } d(\mathrm{Cen}(G), \mathrm{Med}(G)) = k.$$

5.4 The Periphery of a Graph

A vertex v in a connected graph G is a *peripheral vertex* of G if $e(v) = \mathrm{diam}(G)$. The subgraph induced by the peripheral vertices of a connected graph G is the *periphery* of G and is denoted by $\mathrm{Per}(G)$. While every graph can be the center of some graph, Bielak and Syslo showed that only certain graphs can be the periphery of a graph.

Theorem 5.4.1 ([1]) *A nontrivial graph H is the periphery of some graph if and only if every vertex of H has eccentricity 1 or no vertex of H has eccentricity 1.*

Proof If every vertex of H has eccentricity 1, then H is complete and $\mathrm{Per}(H) = H$. Suppose that no vertex of H has eccentricity 1. Let F be the graph obtained from H by adding a new vertex w and joining w to each vertex of H. Since $e_F(w) = 1$ and $e_F(x) = 2$ for every vertex x of H, it follows that every vertex of H is a peripheral vertex of F and so $\mathrm{Per}(F) = F[V(H)] = H$.

For the converse, assume, to the contrary, that there exists a graph H where some vertices of H have eccentricity 1 and some vertices of H have eccentricity greater than 1 for which there is a graph G such that $\mathrm{Per}(G) = H$. Necessarily, H is a proper induced connected subgraph of G and $\mathrm{diam}(G) = k \geq 2$. Furthermore, $e_G(v) = k \geq 2$ for each $v \in V(H)$ and $e_G(v) < k$ for $v \in V(G) - V(H)$. Let u be a vertex of H such that $e_H(u) = 1$ and let w be a vertex of G such that $d_G(u, w) = e_G(u) = k \geq 2$. Since w is not adjacent to u, it follows that $w \notin V(H)$. On the other hand, $d_G(u, w) = k$ and so $e_G(w) = k$. This implies that w is a peripheral vertex of G and so $w \in V(H)$, which is impossible.   ∎

The graph G of Fig. 5.5 has radius 2 and diameter 3. Therefore, every vertex of G is either a central vertex or a peripheral vertex. Indeed, the center of G is the triangle induced by the three vertices v_1, v_2, v_3 of G lying on the outer 3-cycle, while the periphery of G is the 6-cycle induced by the six $u_1, u_2, \ldots, u_6$ vertices of G lying on the inner 6-cycle.

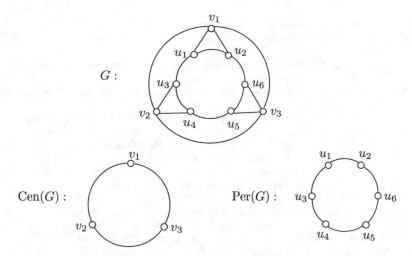

Fig. 5.5 The center and periphery of a graph

A problem involving centers and peripheries is the following.

Problem 5.4.2 Let H_1 be a graph and let H_2 be a graph each of whose vertices has eccentricity 1 or each of whose vertices has eccentricity 2 or more. Does there exist a connected graph G such that

$$\text{Cen}(G) \cong H_1 \text{ and } \text{Per}(G) \cong H_2$$

There is another distance-defined subgraph of a connected graph that does not appear to have been studied extensively. A vertex v in a connected graph G is called an *exterior vertex* of G if v has the maximum total distance in G. The *exterior* $\text{Ext}(G)$ of G is the subgraph induced by the exterior vertices of G.

Conjecture 5.4.3 *For every graph H, there exists a connected graph G such that*

$$\text{Ext}(G) \cong H.$$

Problem 5.4.4 Investigate, for a connected graph G, the relationship between the exterior and the periphery of G.

References

1. H. Bielak, M.M. Syslo, Peripheral vertices in graphs. Stud. Sci. Math. Hung. **18**, 269–275 (1983)
2. F. Buckley, F. Harary, *Distance in Graphs* (Addison-Wesley, Redwood City, 1990)
3. F. Buckley, Z. Miller, P.J. Slater, On graphs containing a given graph as center. J. Graph Theory **5**, 427–434 (1981)

4. F. Harary, Status and contrastatus. Sociometry **22**, 23–43 (1959)
5. F. Harary, R.Z. Norman, The dissimilarity characteristic of Husimi trees. Ann. Math. **58**, 134–141 (1953)
6. S.M. Hedetniemi, S.T. Hedetniemi, P.J. Slater, Centers and medians of $C_{(N)}$-trees. Util. Math. **21**, 225–234 (1982)
7. K.S. Holbert, A note on graphs with distant center and median, in *Recent Studies in Graph Theory*, ed. by V.R. Kulli (Vishna, Gulbarza, 1989), pp. 155–158
8. C. Jordan, Sur les assemblages des lignes. J. Reine Anew. Math. **70**, 185–190 (1869)
9. Z. Miller, Median and degree sequences in graphs. Ars Comb. **15**, 169–177 (1983)
10. S.L. Mitchell, E.J. Cockayne, S.T. Hedetniemi, Linear algorithms for finding the Jordan centre and path centre of a tree. Trans. Sci. **15**, 98–114 (1981)
11. K.S. Novotny, S. Tian, On graphs with intersecting center and median, in *Advances in Graph Theory*, ed. by V.R. Kulli (Vishwa, Gulbarza, 1991), pp. 297–300
12. P.A. Ostrand, Graphs with specified radius and diameter. Discret. Math. **4**, 71–75 (1973)
13. A. Proskurowski, Centers of maximal outerplanar graphs. J. Graph Theory **4**, 75–79 (1980)
14. A. Proskurowski, Centers of 2-trees. Ann. Discret. Math. **9**, 1–5 (1980)
15. P.J. Slater, Medians of arbitrary graphs. J. Graph Theory **4**, 389–392 (1980)
16. P.J. Slater, Locating central paths in a graph. Trans. Sci. **16**, 1–18 (1981)
17. M. Truszyński, Centers and centroids of unicyclic graphs. Math. Slovaca. **35**, 223–228 (1985)

Chapter 6
Eulerian and Hamiltonian Walks

There are two topics in graph theory with a long history, both of which involve traversing graphs, one traversing all the edges in a graph and the second traversing all the vertices in a graph. It is these two topics that are discussed in this chapter.

6.1 Eulerian Walks

A *walk* in a graph G is a sequence of vertices in G such that consecutive vertices are adjacent in G. The number of edges encountered in a walk (including multiplicities) is the *length* of the walk. A walk whose initial and terminal vertices are distinct is an *open walk*; otherwise, it is a *closed walk*. A *trail* is a walk in which no edge is repeated, while a *circuit* is a nontrivial closed trail.

A circuit C in a connected graph G is *Eulerian* if C contains every edge of G, while an open trail T in G is *Eulerian* if T contains every edge of G. A connected graph G is *Eulerian* if G contains an Eulerian circuit. In a famous 1736 paper [4] by Leonhard Euler in which he solved the well-known *Königsberg Bridge Problem*, Euler stated (in graph theory terminology) that a nontrivial connected graph G is Eulerian if and only if every vertex of G has even degree, while G has an Eulerian trail if and only if G has exactly two odd vertices. In his paper, Euler proved that if G is Eulerian, then every vertex of G has even degree. His argument for the converse, however, was considered incomplete and a proof of this converse didn't appear until 1873, in a paper [9] by Hierholzer, which was published two years after his death. Eulerian trails, circuits, and graphs are, of course, named for Euler. The following result is referred to as Euler's theorem.

Theorem 6.1.1 (Euler's Theorem) *A nontrivial connected graph G is Eulerian if and only if every vertex of G has even degree.*

G. Chartrand et al., *From Domination to Coloring*, SpringerBriefs in Mathematics, https://doi.org/10.1007/978-3-030-31110-0_6

It is therefore a consequence of Euler's theorem that if G is a connected graph containing odd vertices, then G fails to contain a closed walk traversing every edge of G exactly once. Consequently, if G is a connected graph containing odd vertices and a closed walk that traverses every edge of G, then some edge of G must be traversed more than once. It is this observation that is the basis for the principal concept of this section.

Let G be a nontrivial connected graph of size m. An *Eulerian walk* in G is a closed walk of minimum length that traverses every edge of G at least once. The length of an Eulerian walk in G is called the *Eulerian number* of G and is denoted by $e(G)$. Clearly, $e(G) \geq m$ and $e(G) = m$ if and only if G is Eulerian. Furthermore, if we replace each edge of G by two parallel edges, then we obtain an Eulerian multigraph M. An Eulerian circuit of M gives rise to a closed walk of G traversing every edge of G twice. Therefore, $e(G) \leq 2m$. The primary problem here is to determine $e(G)$ for an arbitrary connected graph G.

This problem was discussed in 1968 and solved by Stephen Hedetniemi [8] for all trees and connected graphs containing exactly two odd vertices. In 1973, however, Hedetniemi and Goodman [5] solved the more general problem in terms of another graphical parameter. This problem was looked at even earlier (in 1962) by the Chinese mathematician Mei-Ko Kwan [10] and became known as the *Chinese Postman Problem*. Suppose that a postman starts from the post office and has mail to deliver to the houses along each street on his mail route. Once he has completed delivering the mail, he returns to the post office. The problem is to find the minimum length of a round trip that accomplishes this, as we state next.

The Chinese Postman Problem *Determine the minimum length of a round trip that traverses every road in a mail route at least once.*

In graph theory terminology, the Chinese Postman Problem is to determine the length of Eulerian walk in a connected graph. The result corresponding to Theorem 6.1.1 for Eulerian trails (also stated, using graph theory terminology, by Euler) is the following.

Theorem 6.1.2 ([4]) *A nontrivial connected graph G has an Eulerian trail if and only if G contains exactly two vertices of odd degree. Any Eulerian trail in G then begins at one of these odd vertices and terminates at the other.*

By Theorems 6.1.1 and 6.1.2, it therefore follows that if G is any connected graph containing more than two odd vertices, then G contains neither an Eulerian circuit nor an Eulerian trail. Chartrand, Polimeni, and Stewart established the following result.

Theorem 6.1.3 ([1]) *If G is a connected graph containing $2k \geq 4$ odd vertices, then G can be decomposed into k open trails connecting odd vertices, at most one trail of which has odd length.*

We have already noted that for every nontrivial connected graph G, there is always a closed walk traversing every edge of G twice. For an Eulerian walk in G, no edge of G need be traversed more than twice.

Proposition 6.1.4 *Let G be a nontrivial connected graph. No edge of G appears more than twice in any Eulerian walk of G.*

Proof   Assume, to the contrary, that there is an Eulerian walk W of G such that some edge $e = uv$ of G appears three (or more) times on W. There are three possibilities for the location of e in W.

Case 1. $W = W_1, u, v, W_2, u, v, W_3, u, v, W_4$ where W_i is a subwalk (*possibly empty*) of W for $1 \leq i \leq 4$. Let W_2' denote the walk W_2 traversed in reverse order. Then $W^* = W_1, u, W_2', v, W_3, u, v, W_4$ is a closed walk traversing every edge of G whose length is less than that of W. This contradicts the assumption that W is an Eulerian walk of G.

Case 2. $W = W_1, u, v, W_2, u, v, W_3, v, u, W_4$. Here, $W^* = W_1, u, W_2', u, W_3, v, u, W_4$ is a closed walk traversing every edge of G whose length is less than that of W. This is impossible since W is an Eulerian walk of G.

Case 3. $W = W_1, u, v, W_2, v, u, W_3, u, v, W_4$. In this case, $W^* = W_1, u, W_3, u, v, W_2, v, W_4$ is a closed walk traversing every edge of G whose length is less than that of W. This is a contradiction.   ∎

In fact, Kwan presented the following characterization of Eulerian walks.

Theorem 6.1.5 ([10]) *Let G be a nontrivial connected graph. A closed walk W of G traversing every edge of G is an Eulerian walk if and only if*

(1) no edge of G appears more than twice in W and
(2) for each cycle C of G, the number of edges of C that appear twice in W is at most half the length of C.

In particular, Theorem 6.1.5 implies that if G is a connected non-Eulerian graph and W is an Eulerian walk in G, then the subgraph of G induced by the edges appearing twice in W contains no cycles and is therefore a forest.

Let G be a connected graph of size m containing $2k \geq 4$ odd vertices and let S be the set of the $2k$ odd vertices of G. By a *pair partition* of S is meant a partition of S into k pairs of vertices of S. Thus, a pair partition of S can be expressed as

$$\pi = \{\{u_{11}, u_{12}\}, \{u_{21}, u_{22}\}, \ldots, \{u_{k1}, u_{k2}\}\}$$

where then $\{u_{11}, u_{12}, u_{21}, u_{22}, \ldots, u_{k1}, u_{k2}\} = S$. For a given pair partition π of S, the number $d(\pi)$ represents the sum

$$d(\pi) = \sum_{i=1}^{k} d(u_{i1}, u_{i2}).$$

The number $\mu(G)$ is then defined as

$$\mu(G) = \min\{d(\pi) : \ \pi \text{ is a pair partition of } S\};$$

that is, $\mu(G)$ is the smallest value of $d(\pi)$ over all $(2k)!/2^k$ pair partitions π of S.

Fig. 6.1 A graph G with
four odd vertices

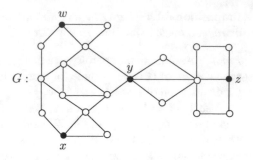

As an illustration, consider the graph G of Fig. 6.1. The set of odd vertices of G is $S = \{w, x, y, z\}$. Here,

$$d(w, x) = 4, d(w, y) = 2, d(w, z) = 4,$$
$$d(x, y) = 3, d(x, z) = 5, d(y, z) = 2.$$

There are three pair partitions of S, namely

$$\pi_1 = \{\{w, x\}, \{y, z\}\}, \pi_2 = \{\{w, y\}, \{x, z\}\}, \pi_3 = \{\{w, z\}, \{x, y\}\}.$$

Since $d(\pi_1) = 6, d(\pi_2) = 7$, and $d(\pi_3) = 7$, it follows that $\mu(G) = 6$.

Goodman and Hedetniemi presented a formula for the Eulerian number of a graph G in terms of its size and the number $\mu(G)$.

Theorem 6.1.6 ([5]) *If G is a connected graph of size m containing at least four odd vertices, then*

$$e(G) = m + \mu(G).$$

Proof To establish the inequality $e(G) \leq m + \mu(G)$, it suffices to show that G contains a closed walk of length $m + \mu(G)$ traversing every edge of G. Suppose that G contains $2k \geq 4$ odd vertices. Let $\pi = \{\{u_{11}, u_{12}\}, \{u_{21}, u_{22}\}, \ldots, \{u_{k1}, u_{k2}\}\}$ be a pair partition of the set S of odd vertices of G such that $\mu(G) = d(\pi)$ and let P_i be a $u_{i1} - u_{i2}$ geodesic of G for $1 \leq i \leq k$. Now, let H be the multigraph obtained from G by replacing each edge of every path P_i $(1 \leq i \leq k)$ by two parallel edges. Thus, H is an Eulerian multigraph of size $m + \mu(G)$. An Eulerian circuit of H produces a closed walk of length $m + \mu(G)$ traversing every edge of G. Thus, $e(G) \leq m + \mu(G)$.

It remains to show that $e(G) \geq m + \mu(G)$. Let W be an Eulerian walk of G. Therefore, the length of W is $e(G)$ and every edge of G appears on W at least once and at most twice by Proposition 6.1.4. Let H be the subgraph of G induced by those edges appearing twice in W. Therefore, $e(G) = m + |E(H)|$. By Theorem 6.1.5, H is a forest. For an even vertex u in G, either u does not belong to H or there is an even number of edges incident with u that belong to H. For an odd vertex u in G, there is an odd number of edges incident with u that belong to H. Hence, every odd vertex of G belongs to H and the set of odd vertices of H is the set S of odd vertices of G. We consider two cases, according to whether H is connected or H is disconnected.

Case 1. *H is connected*. Since H is a tree containing $2k \geq 4$ odd vertices, H can be decomposed into k open trails (paths in this case) $P_1, P_2, \ldots, P_k$ connecting odd vertices (by Theorem 6.1.3). Suppose that P_i is a $u_{i1} - u_{i2}$ path of length ℓ_i for $1 \leq i \leq k$. Then

$$\pi = \{\{u_{11}, u_{12}\}, \{u_{21}, u_{22}\}, \ldots, \{u_{k1}, u_{k2}\}\}$$

is a pair partition of S. Since $d_G(u_{i1}, u_{i2}) \leq \ell_i$ for each integer i with $1 \leq i \leq k$, it follows that

$$|E(H)| = \sum_{i=1}^{k} \ell_i \geq \sum_{i=1}^{k} d_G(u_{i1}, u_{i2}) = d(\pi) \geq \mu(G).$$

Case 2. *H is disconnected*. Then H is a forest consisting of $p \geq 2$ components $H_1, H_2, \ldots, H_p$. Therefore, each component H_i is a nontrivial tree. For $1 \leq i \leq p$, let m_i be the size of H_i and let k_i be the number of odd vertices of H_i. Then $k_i \geq 2$ is even for $1 \leq i \leq p$. For $1 \leq i \leq p$, let S_i be the set of all odd vertices of H_i. Then $\{S_1, S_2, \ldots, S_p\}$ is a partition of the set S of odd vertices of G. Applying the same argument used in Case 1 to each component H_i for $1 \leq i \leq p$, we see that for each integer i with $1 \leq i \leq p$, there is a pair partition π_i of S_i such that $m_i \geq d(\pi_i)$. Since $\{S_1, S_2, \ldots, S_p\}$ is a partition of S, it follows that $\pi_1, \pi_2, \ldots, \pi_p$ gives rise to a pair partition π of S such that

$$d(\pi) = \sum_{i=1}^{p} d(\pi_i).$$

Therefore,

$$|E(H)| = \sum_{i=1}^{p} m_i \geq \sum_{i=1}^{p} d(\pi_i) = d(\pi) \geq \mu(G).$$

Consequently, $e(G) = m + |E(H)| \geq m + \mu(G)$, completing the proof. ∎

6.2 Hamiltonian Walks

In 1856, Sir William Rowan Hamilton developed a game he called the *Icosian Game*. One form of this game consisted of a board on which was a diagram of the graph of the dodecahedron. The twenty vertices of this diagram were designated by the twenty consonants of the English alphabet. In one version of the game, called "Around the World", each vertex corresponded to a world city whose name began with the corresponding consonant and the player was to find a round trip that passed through every city exactly once. In fact, expressed in graph theory terms, Hamilton stated that if one would begin with any path of order 5 on this graph, a cycle could always be found on the graph that contained this path as well as all twenty vertices. This eventually led to a popular concept named for Hamilton.

A cycle in a graph G containing every vertex of G is called a *Hamiltonian cycle* of G, while a graph possessing a Hamiltonian cycle is a *Hamiltonian graph*. Unlike the situation for Eulerian graphs, there is no corresponding characterization of Hamiltonian graphs. There are, however, several sufficient conditions for a graph to be Hamiltonian, many of which require the graph to have diameter at most 2. The first of these results was obtained by Dirac in 1952.

Theorem 6.2.1 ([3]) *If G is a graph of order $n \geq 3$ such that $\delta(G) \geq n/2$, then G is Hamiltonian.*

Eight years later, in 1960, Ore obtained a more general sufficient condition for a graph to be Hamiltonian. For a noncomplete graph G, let $\sigma_2(G)$ denote the minimum degree sum of every two nonadjacent vertices of G.

Theorem 6.2.2 ([11]) *If G is a graph of order $n \geq 3$ such that $\sigma_2(G) \geq n$, then G is Hamiltonian.*

One consequence of Theorem 6.2.2 gives a sufficient condition for a graph to be Hamiltonian in terms of its size.

Corollary 6.2.3 *If G is a graph of order $n \geq 3$ and size $m \geq \binom{n-1}{2} + 2$, then G is Hamiltonian.*

While all of the bounds in these results are sharp, there are many Hamiltonian graphs satisfying none of these bounds. If a graph G is not Hamiltonian, then, by definition, there is no cycle in G containing all vertices of G. On the other hand, for every nontrivial connected graph G, there are closed walks in G traversing every vertex of G. In fact, every Eulerian walk in G has this property.

Let G be a nontrivial connected graph. A closed walk of minimum length containing every vertex of G is a *Hamiltonian walk* and the length of a Hamiltonian walk in G is called the *Hamiltonian number* of G, denoted by $h(G)$. Therefore, if G is a connected graph of order $n \geq 2$, then $h(G) \geq n$. Furthermore, $h(G) = n$ if and only if G is Hamiltonian. These concepts were introduced by Stephen Hedetniemi along with Seymour Goodman [6, 7], who established the following bounds for the Hamiltonian number of a graph in terms of its order, size, and Eulerian number.

Proposition 6.2.4 ([6, 7]) *If G is a connected graph of order $n \geq 2$ and size m, then*

$$n \leq h(G) \leq e(G) \leq 2m.$$

Proof   It has already been observed that $h(G) \geq n$ and that replacing each edge of G by two parallel edges produces an Eulerian multigraph and so $e(G) \leq 2m$. Furthermore, every Eulerian walk in G is also a Hamiltonian walk of G. These observations produce the desired inequalities.   ∎

Since every bridge in a nontrivial connected graph G must be traversed more than once in a Hamiltonian walk of G, the following result is immediate.

Proposition 6.2.5 *If T is a nontrivial tree of order n, then*

$$h(T) = 2n - 2.$$

By Proposition 6.1.4, no edge of a nontrivial connected graph G appears more than twice in any Eulerian walk of G. This is also the case for Hamiltonian walks of G.

Proposition 6.2.6 ([7]) *Every edge of a nontrivial connected graph G appears at most twice in any Hamiltonian walk of G.*

Goodman and Hedetniemi provided a necessary condition for a nontrivial connected graph to have the same Eulerian and Hamiltonian numbers.

Theorem 6.2.7 ([7]) *Let G be a nontrivial connected graph. If $h(G) = e(G)$, then for every cycle C of G, the spanning subgraph $G - E(C)$ is disconnected.*

Proof Assume, to the contrary, that there is a nontrivial connected graph G for which $h(G) = e(G)$, but G contains a cycle C, of length k say, such that $H = G - E(C)$ is connected. Let W be an Eulerian walk of G and let M be the multigraph obtained from G corresponding to W. Hence, the size of M is $e(G)$. Let F be the multigraph obtained by deleting every edge of C from M. Since F is connected, F is Eulerian. However then,

$$h(G) \le e(F) = |E(F)| = e(G) - k < e(G),$$

which is a contradiction. ∎

That the Eulerian number of a connected non-Eulerian graph G can be computed by determining all pair partitions of the set of odd vertices of G and minimizing the sum of the distances of the vertices in each pair in such a partition suggested another concept introduced by Chartrand, Thomas, Saenpholphat, and Zhang in [2], which has a direct connection with the Hamiltonian number of a connected graph.

For every Hamiltonian cycle $(v_1, v_2, \ldots, v_n, v_{n+1} = v_1)$ of a Hamiltonian graph G, it follows that $v_i v_{i+1} \in E(G)$ for $1 \le i \le n$. Hamiltonian graphs of order $n > 3$ are therefore those graphs for which there is a cyclic ordering $v_1, v_2, \ldots, v_n, v_{n+1} = v_1$ of the vertices of G such that $\sum_{i=1}^{n} d(v_i, v_{i+1}) = n$, where $d(v_i, v_{i+1})$ is the distance between v_i and v_{i+1} for $1 \le i \le n$. For a connected graph G of order $n \ge 3$ and a cyclic ordering $s : v_1, v_2, \ldots, v_n, v_{n+1} = v_1$ of the vertices of G, the number $d(s)$ is defined by

$$d(s) = \sum_{i=1}^{n} d(v_i, v_{i+1}).$$

Fig. 6.2 A graph G with
$h^*(G) = 6$

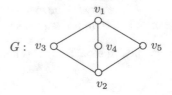

$$G: \quad v_3$$

Therefore, $d(s) \geq n$ for each cyclic ordering s of the vertices of G. The number $h^*(G)$ is defined by

$$h^*(G) = \min \{d(s) : s \text{ is a cyclic ordering of the vertices of } G\}.$$

Consider the graph G of Fig. 6.2. For the cyclic orderings

$$s_1 : v_1, v_2, v_3, v_4, v_5, v_1 \text{ and } s_2 : v_1, v_3, v_2, v_4, v_5, v_1$$

of the vertices of G, we see that $d(s_1) = 8$ and $d(s_2) = 6$. Since G is a non-Hamiltonian graph of order 5 and $d(s_2) = 6$, it follows that $h^*(G) = 6$.

It was shown in [2] that there is an alternative way to define the Hamiltonian number $h(G)$ of a nontrivial connected graph G. The *length* of a walk W in a graph is denoted by $L(W)$.

Theorem 6.2.8 ([2]) *For every connected graph G,*

$$h^*(G) = h(G).$$

Proof To show that $h(G) \leq h^*(G)$, let $s : v_1, v_2, \ldots, v_n, v_{n+1} = v_1$ be a cyclic ordering of the vertices of G for which $d(s) = h^*(G)$. For each integer i with $1 \leq i \leq n$, let P_i be a $v_i - v_{i+1}$ geodesic in G. Thus, $L(P_i) = d(v_i, v_{i+1})$. The union of the paths P_i forms a closed spanning walk W in G. Therefore,

$$h(G) \leq L(W) = \sum_{i=1}^{n} L(P_i) = \sum_{i=1}^{n} d(v_i, v_{i+1}) = d(s) = h^*(G).$$

To show that $h^*(G) \leq h(G)$, let W be a Hamiltonian walk in G. Therefore, $L(W) = h(G)$. Suppose that $W = (x_1, x_2, \ldots, x_N, x_1)$, where then $N \geq n$. Define $v_1 = x_1$ and $v_2 = x_2$. For $3 \leq i \leq n$, define v_i to be x_{j_i}, where j_i is the smallest positive integer such that $x_{j_i} \notin \{v_1, v_2, \ldots, v_{i-1}\}$. Then $s : v_1, v_2, \ldots, v_n, v_{n+1} = v_1$ is a cyclic ordering of the vertices of G. For each integer i with $1 \leq i \leq n$, let W_i be the $v_i - v_{i+1}$ subwalk of W and so $d(v_i, v_{i+1}) \leq L(W_i)$. Since

$$h^*(G) \leq \sum_{i=1}^{n} d(v_i, v_{i+1}) \leq \sum_{i=1}^{n} L(W_i) = L(W) = h(G),$$

we have the desired result.   ∎

As a consequence of Theorem 6.2.8, the number $h^*(G)$ for a nontrivial connected graph G is the Hamiltonian number of G, which we henceforth denote by $h(G)$. That is, $h^*(G) = h(G)$ is the length of a Hamiltonian walk in G.

By Proposition 6.2.5, if T is a tree of order n, then $h(T) = 2n - 2$. To show that the converse of this statement holds as well, the following lemma is useful.

Lemma 6.2.9 ([2]) *If G is a connected graph such that $\delta(G) \geq 2$ and $\Delta(G) \geq 3$, then G contains two distinct cycles C and C' such that $V(C) \neq V(C')$.*

Theorem 6.2.10 ([2]) *If G is a nontrivial connected graph of order n, then*

$$h(G) = 2n - 2 \text{ if and only if } G \text{ is a tree.}$$

Proof By Proposition 6.2.5, it suffices to show that if G is a connected graph of order $n \geq 3$ that is not a tree, then $h(G) < 2n-2$. We proceed by induction on n. Since $h(K_3) = 3$, the result holds for $n = 3$. Suppose that $h(F) < 2(n - 1) - 2 = 2n - 4$ for all connected graphs F of order $n - 1 \geq 3$ that are not trees. Let G be a connected graph of order $n \geq 4$ that is not a tree. Since $h(C_n) = n < 2n - 2$, we may assume that $G \neq C_n$.

We claim that G contains a vertex u such that $G - u$ is a connected subgraph of G that is not a tree. If G contains cut-vertices, then there is a vertex u that is a non-cut-vertex of an end-block that has the desired property. So, we may assume that G is 2-connected and $\delta(G) \geq 2$. By Lemma 6.2.9, G contains two distinct cycles C and C' with $V(C) \neq V(C')$. Thus, if u is a vertex that belongs to one of C and C' but not the other, then $G - u$ is a connected subgraph of G that is not a tree. By the induction hypothesis, $h(G - u) < 2(n - 1) - 2 = 2n - 4$. Let

$$s_0 : v_1, v_2, \ldots, v_{n-1}, v_1$$

be a cyclic ordering of the vertices of $G - u$ with $d(s_0) = h(G - u) < 2n - 4$. Suppose that u is adjacent to the vertex v_i, where $1 \leq i \leq n - 1$. Define the cyclic ordering s_0' of the vertices of G from s_0 by

$$s_0' : v_1, v_2, \ldots, v_i, u, v_{i+1}, \ldots, v_{n-1}, v_1.$$

Since $d(v_i, u) = 1$, it follows by the triangle inequality that

$$d(u, v_{i+1}) \leq 1 + d(v_i, v_{i+1}).$$

Therefore,

$$\begin{aligned}
d(s_0') &= d(s_0) - d(v_i, v_{i+1}) + d(v_i, u) + d(u, v_{i+1}) \\
&\leq d(s_0) - d(v_i, v_{i+1}) + 1 + [1 + d(v_i, v_{i+1})] \\
&= d(s_0) + 2 < (2n - 4) + 2 = 2n - 2.
\end{aligned}$$

Therefore, $h(G) \leq d(s_0') < 2n - 2$, as desired. ∎

For the graph G of Fig. 6.2, it was seen that there are cyclic orderings s_1 and s_2 of the vertices of G such that $d(s_1) = 8$ and $d(s_2) = 6$. Indeed, it is not difficult to see that for *every* cyclic ordering s of $V(G)$, either $d(s) = 6$ or $d(s) = 8$.

For a connected graph G, the *upper Hamiltonian number* $h^+(G)$ is defined by

$$h^+(G) = \max \{d(s) : s \text{ is a cyclic ordering of the vertices of } G\}.$$

From the remarks above, it follows that $h^+(G) = 8$ for the graph G of Fig. 6.2. Obviously,

$$h(G) \leq h^+(G)$$

for every connected graph G. For each integer $n \geq 3$, there are only two connected graphs G of order n for which $h(G) = h^+(G)$.

Theorem 6.2.11 ([2]) *If G is a connected graph of order $n \geq 3$, then*

$$h(G) = h^+(G) \text{ if and only if } G = K_n \text{ or } G = K_{1,n-1}.$$

Proof If $G = K_n$, then certainly $d(s) = n$ for every cyclic ordering s of the vertices of G; while if $G = K_{1,n-1}$, then $d(s) = 2n - 2$ for every cyclic ordering s of the vertices of G. Thus, $h(G) = h^+(G)$ if $G = K_n$ or $G = K_{1,n-1}$.

For the converse, suppose that G is a connected graph of order $n \geq 3$ such that $G \neq K_n, K_{1,n-1}$. We show that $h(G) \neq h^+(G)$. Let diam $G = d$. Since $G \neq K_n$, it follows that $d \geq 2$. Therefore, either $d \geq 3$ or $d = 2$.

Case 1. $d \geq 3$. Let v_1 and v_{d+1} be vertices of G such that $d(v_1, v_{d+1}) = d$ and let $P = (v_1, v_2, \ldots, v_{d+1})$ be a $v_1 - v_{d+1}$ geodesic in G. Let $W = V(G) \setminus V(P)$. If $W \neq \emptyset$, then let $W = \{w_1, w_2, \ldots, w_\ell\}$, where $\ell = n - d - 1$. Define a cyclic ordering s of the vertices of G by

$$s : v_1, v_2, v_3, \ldots, v_{d+1}, v_1 \text{ or} \tag{6.1}$$

$$s : v_1, v_2, v_3, \ldots, v_{d+1}, w_1, w_2, \ldots, w_\ell, v_1, \tag{6.2}$$

according to whether $W = \emptyset$ or $W \neq \emptyset$. Let s' be the cyclic ordering of the vertices of G obtained from s by interchanging the locations of v_2 and v_3 in s; that is,

$$s' : v_1, v_3, v_2, v_4, \ldots, v_{d+1}, v_1 \tag{6.3}$$

$$\text{or } s' : v_1, v_3, v_2, v_4, \ldots, v_{d+1}, w_1, w_2, \ldots, w_\ell, v_1, \tag{6.4}$$

according to whether $W = \emptyset$ or $W \neq \emptyset$. In either case, $d(s') = d(s) + 2$ and so $h(G) \neq h^+(G)$.

Case 2. $d = 2$. Since G is not a star, it follows that G is not a tree. Thus, the girth $g(G) = k \geq 3$. Assume first that $k = 3$. Since G is connected and $G \neq K_n$, there exists a set U of four vertices of G such that $G[U] = K_4 - e$ or $G[U]$ is a triangle

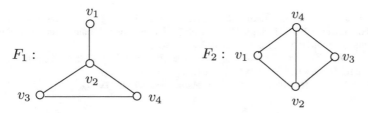

Fig. 6.3 Induced subgraphs F_1 and F_2 of G in Case 2

with a pendant edge. Therefore, we may assume, without loss of generality, that G contains one of the graphs F_1 and F_2 in Fig. 6.3 as an induced subgraph. In either case, define the cyclic orderings s and s' as described in (6.1) (or (6.2)) and (6.3) (or (6.4)), respectively. Then $d(s') = d(s) + 1$ and so $h(G) \neq h^+(G)$.

If $k \geq 4$, then let $C = (v_1, v_2, \ldots, v_k, v_1)$ be an induced cycle of G and let $V(G) - V(C) = \{w_1, w_2, \ldots, w_\ell\}$ if $\ell = n - k > 0$. Define the cyclic orderings s and s' of the vertices of G as in (6.1) (or (6.2)) and (6.3) (or (6.4)), respectively. Since $d(s') = d(s) + 2$, it follows that $h(G) \neq h^+(G)$. ∎

Problem 6.2.12 It was shown in [2] that if T is a tree of order $n \geq 3$, then

$$2n - 2 \leq h^+(T) \leq \left\lfloor n^2/2 \right\rfloor.$$

Furthermore,
$$h^+(T) = 2n - 2 \ if \ and \ only \ if \ T = K_{1,n-1},$$
$$and$$
$$h^+(T) = \left\lfloor n^2/2 \right\rfloor \ if \ and \ only \ if \ T = P_n.$$

What other information can be obtained on the upper Hamiltonian numbers of trees?

References

1. G. Chartrand, A.D. Polimeni, M.J. Stewart, The existence of 1-factors in line graphs, squares, and total graphs. Nedrl. Akad. Wetensch. Proc. Ser. A **76**; Indag. Math. **35**, 228–232 (1973)
2. G. Chartrand, T. Thomas, V. Saenpholphat, P. Zhang, A new look at Hamiltonian walks. Bull. Inst. Comb. Appl. **42**, 37–52 (2004)
3. G.A. Dirac, Some theorems on abstract graphs. Proc. Lond. Math. Soc. **2**, 69–81 (1952)
4. L. Euler, Solutio problematis ad geometriam situs pertinentis. Comment. Acad. Sci. Imp. Petropol. **8**, 128–140 (1736)
5. S.E. Goodman, S.T. Hedetniemi, Eulerian walks in graphs. SIAM J. Comput. **2**, 16–27 (1973)
6. S.E. Goodman, S.T. Hedetniemi, On Hamiltonian walks in graphs. Congr. Numer. 335–342 (1973)
7. S.E. Goodman, S.T. Hedetniemi, On Hamiltonian walks in graphs. SIAM J. Comput. **3**, 214–221 (1974)

8. S.T. Hedetniemi, On minimum walks in graphs. Nav. Res. Logist. Q. **15**, 453–458 (1968)
9. C. Hierholzer, Über die Möglichkeit, einen Linienzug ohne Wiederholung und ohne Unter-brechnung zu umfahren. Math. Ann. **6**, 30–32 (1873)
10. M.K. Kwan, Graphic programming using odd or even points. Acta Math. Sin. **10**, 264–266 (Chinese) (1960); translated as Chin. Math. **1**, 273–277 (1960)
11. O. Ore, Note on Hamilton circuits. Am. Math. Mon. **67**, 55 (1960)

Chapter 7
Complete Colorings

In a proper k-coloring of a k-chromatic graph, for every two distinct colors there are always adjacent vertices with these colors. This observation has led to a coloring called a complete coloring, which is the primary topic of this chapter. We investigate the largest number of colors required of a complete coloring as well as several related coloring parameters.

7.1 Introduction

One of the best known topics in graph theory is graph colorings, an area whose history goes back well over a century. The interest in graph colorings grew out of the many attempts to solve the famous Four Color Problem, introduced by Francis Guthrie in 1852. While graph colorings have been looked at in many ways in recent decades, the best known and most studied are proper colorings, both proper vertex colorings and proper edge colorings. In a *proper vertex coloring*, every two adjacent vertices are required to be colored differently. A proper vertex coloring whose colors are taken from a set of k colors, usually the set $[k] = \{1, 2, \ldots, k\}$, is called a *proper k-coloring*. The parameter of greatest interest here is the *chromatic number* of a graph G, denoted $\chi(G)$, and defined as the smallest positive integer k for which G has a proper k-coloring. If $\chi(G) = k$, then G is referred to as a *k-chromatic graph*.

If a graph G has chromatic number k, then for every proper k-coloring of the vertices of G and for every two distinct colors $i, j \in [k]$, there are adjacent vertices of G, one colored i and the other colored j. (If this were not the case, then the set V_i of vertices colored i and the set V_j of vertices colored j could be merged into a single independent set $V_i \cup V_j$ and all vertices in this set could be assigned the same color, resulting in a $(k-1)$-coloring of G, which is impossible.) On the other hand, it is possible that there is a proper k-coloring of a graph H, where all k colors are used and for every two distinct colors, there are adjacent vertices in G assigned these

G. Chartrand et al., *From Domination to Coloring*, SpringerBriefs in Mathematics,
https://doi.org/10.1007/978-3-030-31110-0_7

Fig. 7.1 A proper
5-coloring of P_{11}

colors but yet $\chi(G) \neq k$. For example, the proper 5-coloring of the path P_{11} shown
in Fig. 7.1 has this property and yet $\chi(P_{11}) \neq 5$; indeed, $\chi(P_{11}) = 2$. This leads us
to the main concept of this chapter.

7.2 The Achromatic Number of a Graph

By a *complete coloring* of a graph G is meant a proper vertex coloring of G having
the property that for every two distinct colors i and j used in the coloring, there exist
adjacent vertices of G colored i and j. A complete coloring in which k colors are
used is a *complete k-coloring*. In order for a graph to has a complete k-coloring, it
must have at least $\binom{k}{2}$ edges. As mentioned above, every k-coloring of a k-chromatic
graph is a complete k-coloring. Consequently, the minimum positive integer k for
which a graph G has a complete k-coloring is $\chi(G)$. As the graph P_{11} shows in
Fig. 7.1, it is possible for a graph G to have a complete k-coloring where $k > \chi(G)$.

The largest positive integer k for which G has a complete k-coloring is the *achromatic number* of G, which is denoted $\psi(G)$. Hence, $\psi(G) \geq \chi(G)$ for every graph G.
This parameter was introduced in [5] by Stephen Hedetniemi, his advisor (academic
father) Frank Harary, and Geert Prins (an academic brother of Hedetniemi). If G is
a graph of size m, then

$$\psi(G) \leq \frac{1 + \sqrt{1 + 8m}}{2}.$$

Since the graph P_{11} shown in Fig. 7.1 has size 10 and a complete 5-coloring, it follows
that $\psi(P_{11}) = 5$. The achromatic numbers of all paths and cycles were determined
by Hell and Miller.

Theorem 7.2.1 ([7]) *For each integer $n \geq 2$,*

$$\psi(P_n) = \max \left\{ k : \left(\left\lfloor \frac{k}{2} \right\rfloor + 1 \right) (k - 2) + 2 \leq n \right\}.$$

Theorem 7.2.2 ([7]) *For each integer $n \geq 3$,*

$$\psi(C_n) = \max \left\{ k : k \left\lfloor \frac{k}{2} \right\rfloor \leq n \right\} - s(n),$$

where $s(n)$ is the number of positive integer solutions of $n = 2x^2 + x + 1$.

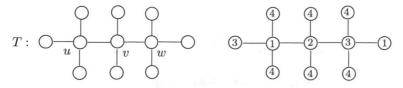

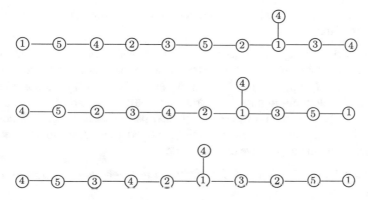

Fig. 7.2 A tree of size 10 with achromatic number 4

Fig. 7.3 Trees of size 10 having achromatic number 5

While the path P_{11} has size 10 and achromatic number 5, this is not true for all trees of size 10. For example, the tree T in Fig. 7.2 has size 10 but $\psi(T) \neq 5$. To see this, suppose that $\psi(T) = 5$. Then there exists a complete 5-coloring of T using the colors in the set [5]. Then there are two colors, say 4 and 5, assigned to none of u, v, and w. However, only end-vertices can be colored 4 and 5 and no two end-vertices are adjacent. Thus, there are no adjacent vertices colored 4 and 5, which is impossible. In fact, $\psi(T) = 4$ as the complete 4-coloring of T in Fig. 7.2 shows.

The path P_{11} is the only tree of size 10 having diameter 10 and, as we saw, $\psi(P_{11}) = 5$. The three trees of size 10 in Fig. 7.3 have diameter 9 and achromatic number 5.

While the three trees in Fig. 7.3 have achromatic number 5, this is not the case for all trees of size 10 and diameter 9. The tree T shown in Fig. 7.4 has diameter 9 and achromatic number 4. To see that $\psi(T) \neq 5$, suppose that there exists a complete 5-coloring c of T using the colors in the set [5]. Then, for each color $i \in [5]$, there is exactly one vertex colored i having a unique neighbor colored $j \in [5]$ with $j \neq i$. We may assume that $c(w) = 1$. Since deg $w = 3$, there is only one other vertex colored 1 and this must be an end-vertex of T. Necessarily, u is colored 1. Suppose that $c(x) = 2$. Then some other end-vertex must be colored 2, which means that v is colored 2. But there are two pairs of vertices colored 1 and 2, which is impossible. Since the tree T has a complete 4-coloring, it follows that $\psi(T) = 4$.

At the other extreme, the star $K_{1,10}$ has size 10, diameter 2, and achromatic number 2. Indeed, every star has achromatic number 2. This brings up the following.

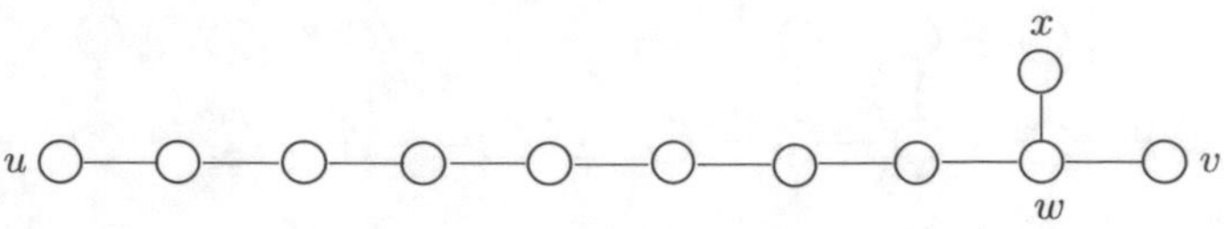

Fig. 7.4 A tree of size 10, diameter 9 and achromatic number 4

Problem 7.2.3 Study the relationship between the achromatic number and other parameters of trees.

Not only does every star have achromatic number 2, every complete bipartite graph has achromatic number 2. To see this, suppose that there exists some complete bipartite graph G with $\psi(G) \geq 3$. Then, in a complete $\psi(G)$-coloring of G, two vertices of G belonging to the same partite set would have to be assigned different colors, say 1 and 2, while no vertex in any other partite set is assigned either color. However then, there are no adjacent vertices colored 1 and 2, which is impossible. By the same argument, we have the following.

Proposition 7.2.4 *If G is a complete multipartite graph, then $\psi(G) = \chi(G)$.*

7.3 Graph Homomorphisms

One of the fundamental concepts in graph theory is that of isomorphism. To recall, an *isomorphism* from a graph G to a graph H is a bijective function $\phi : V(G) \to V(H)$ that maps adjacent vertices in G to adjacent vertices in H and nonadjacent vertices in G to nonadjacent vertices in H. If such a function exists, then G and H are *isomorphic graphs*. There is a related concept that is of particular interest dealing with proper vertex colorings.

A *homomorphism* from a graph G to a graph G' is a function $\phi : V(G) \to V(G')$ that maps adjacent vertices in G to adjacent vertices in G'. If ϕ is a homomorphism from G to G' and u and v are nonadjacent vertices in G, then any of the following is possible: (1) $\phi(u)$ and $\phi(v)$ are nonadjacent, (2) $\phi(u)$ and $\phi(v)$ are adjacent, (3) $\phi(u) = \phi(v)$. The subgraph H of G' whose vertex set $V(H)$ is the image $\phi(V(G))$ of $V(G)$ under ϕ and whose edge set consists of all those edges $u'v'$ in G' such that $\phi(u) = u'$ and $\phi(v) = v'$ for two adjacent vertices u and v of G is called the *homomorphic image* of G under ϕ and is denoted by $\phi(G) = H$. A graph H is a *homomorphic image* of a graph G if $H = \phi(G)$ for some homomorphism ϕ of G. If H is a homomorphic image of a graph G under a homomorphism ϕ from G to a graph G', then ϕ is also a homomorphism from G to H. It is the topic of homomorphisms that Stephen Hedetniemi investigated in his doctoral dissertation, titled *Homomorphisms of Graphs and Automata* [6]. Steve had two doctoral advisors and, therefore, two academic fathers, namely Frank Harary and John Holland. He received his Ph.D. in 1966. Steve earned all three of his degrees (baccalaureate, Master's, and doctoral) from the University of Michigan.

Fig. 7.5 A graph G and a homomorphic image of G

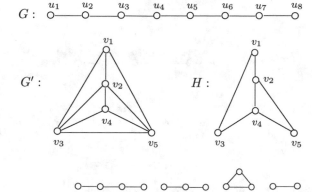

Fig. 7.6 The homomorphic images of P_4

For example, consider the two graphs $G = P_8$ and $G' = K_5 - e$ shown in Fig. 7.5 and the function $\phi : V(G) \to V(G')$ defined by

$$\phi(u_1) = \phi(u_3) = v_1, \phi(u_2) = \phi(u_6) = v_2,$$

$$\phi(u_4) = v_3, \phi(u_5) = \phi(u_8) = v_4, \phi(u_7) = v_5.$$

This function ϕ is a homomorphism from G to G'. The homomorphic image $H = \phi(G)$ of G under ϕ is also shown in Fig. 7.5. Therefore, ϕ is also a homomorphism from G to H.

There are exactly four homomorphic images of the path P_4. These are shown in Fig. 7.6. On the other hand, for each positive integer n, the only homomorphic image of K_n is K_n itself.

There is an alternative way to obtain the homomorphic images of a graph G. As we noted, the only homomorphic image of the complete graph $G = K_n$ is K_n. Otherwise, G is not complete and thus contains one or more pairs of nonadjacent vertices. An *elementary homomorphism* of a graph G is obtained by identifying two nonadjacent vertices u and v of G. The vertex obtained by identifying u and v may be denoted by either u or v. Thus, the resulting homomorphic image G' can be considered to have vertex set $V(G) - \{u\}$ and edge set

$$E(G') = \{xy : xy \in E(G), x, y \in V(G) - \{u, v\}\} \cup$$
$$\{vx : vx \in E(G) \text{ or } ux \in E(G), x \notin V(G) - \{u, v\}\}.$$

Alternatively, the mapping $\epsilon : V(G) \to V(G')$ defined by

$$\epsilon(x) = \begin{cases} x & \text{if } x \in V(G) - \{u, v\} \\ v & \text{if } x \in \{u, v\} \end{cases}$$

is an *elementary homomorphism* from G to G'. The homomorphic image $\epsilon(G)$ of a graph G obtained from an elementary homomorphism ϵ is also referred to as

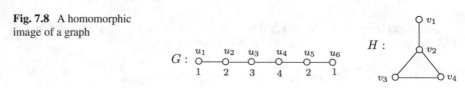

$G:$ $\overset{u_1}{\circ}\!\!-\!\!\overset{u_2}{\circ}\!\!-\!\!\overset{u_3}{\circ}\!\!-\!\!\overset{u_4}{\circ}\!\!-\!\!\overset{u_5}{\circ}\!\!-\!\!\overset{u_6}{\circ}$

$G_1:$

$G_2:$

$H:$

Fig. 7.7 Some homomorphic images of a graph

Fig. 7.8 A homomorphic
image of a graph

$G:$ $\overset{u_1}{\underset{1}{\circ}}\!\!-\!\!\overset{u_2}{\underset{2}{\circ}}\!\!-\!\!\overset{u_3}{\underset{3}{\circ}}\!\!-\!\!\overset{u_4}{\underset{4}{\circ}}\!\!-\!\!\overset{u_5}{\underset{2}{\circ}}\!\!-\!\!\overset{u_6}{\underset{1}{\circ}}$

$H:$

an *elementary homomorphic image*. Not only is G' a homomorphic image of G, a graph H is a homomorphic image of a graph G if and only if H can be obtained by a sequence of elementary homomorphisms beginning with G. For example, if we identify u_1 and u_6 in the graph G of Fig. 7.7, we obtain the graph G_1 shown in Fig. 7.7, which is a homomorphic image of G. Then identifying u_2 and u_5 in G_1, we obtain G_2, which is a homomorphic image of G_1. The graph G_2 is also a homomorphic image of G. The graph G_2 is isomorphic to the graph H also shown in Fig. 7.7.

The fact that each homomorphic image of a graph G can be obtained from G by a sequence of elementary homomorphisms tells us that we can obtain each homomorphic image of G by an appropriate partition $\mathcal{P} = \{V_1, V_2, \ldots, V_k\}$ of $V(G)$ into independent sets such that $V(H) = \{v_1, v_2, \ldots, v_k\}$, where v_i is adjacent to v_j if and only if some vertex in V_i is adjacent to some vertex in V_j. The partition $\mathcal{P}$ of $V(G)$ then corresponds to the coloring c of G in which each vertex in V_i is assigned the color i $(1 \leq i \leq k)$. In particular, if the coloring c is a complete k-coloring, then $H = K_k$. The 4-coloring of the graph G in Fig. 7.7 shown in Fig. 7.8 results in the color classes $V_1 = \{u_1, u_6\}$, $V_2 = \{u_2, u_5\}$, $V_3 = \{u_3\}$, and $V_4 = \{u_4\}$ and the homomorphic image H of Fig. 7.7, which is also shown in Fig. 7.8.

Therefore, if a graph H is a homomorphic image of a graph G, then there is a homomorphism ϕ from G to H and for each vertex v in H, the set $\phi^{-1}(v)$ of those vertices of G having v as their image is independent in G. Consequently, each coloring of H gives rise to a coloring of G by assigning to each vertex of G in $\phi^{-1}(v)$ the color that is assigned to v in H. For this reason, the graph G is said to be H-*colorable*. This provides us with the following observation, which is a primary reason for one's interest in graph homomorphisms.

Theorem 7.3.1 *If H is a homomorphic image of a graph G, then*

$$\chi(G) \le \chi(H).$$

The chromatic number of an elementary homomorphic image of a graph can never exceed the chromatic number of the graph by more than 1.

Theorem 7.3.2 *If ϵ is an elementary homomorphism of a graph G, then*

$$\chi(G) \le \chi(\epsilon(G)) \le \chi(G) + 1.$$

Proof Suppose that ϵ identifies the nonadjacent vertices u and v of G. We have already noted the inequality $\chi(G) \le \chi(\epsilon(G))$). Suppose that $\chi(G) = k$ and a k-coloring of G is given, using the colors $1, 2, \dots, k$. Define a coloring c' of $\epsilon(G)$ by

$$c'(x) = \begin{cases} c(x) & \text{if } x \in V(G) - \{u, v\} \\ k+1 & \text{if } x \in \{u, v\} \end{cases}$$

Since c' is a $(k + 1)$-coloring of $\epsilon(G)$, it follows that

$$\chi(\epsilon(G)) \le k + 1 = \chi(G) + 1,$$

giving the desired result.   ∎

The following theorem is one of the major theorems dealing with graph homomorphisms and is due to Harary, Hedetniemi, and Prins.

Theorem 7.3.3 ([5]) (The Homomorphism Interpolation Theorem) *Let G be a graph. For every integer ℓ with $\chi(G) \le \ell \le \psi(G)$, there is a homomorphic image H of G with $\chi(H) = \ell$.*

Proof The theorem is certainly true if $\ell = \chi(G)$ or $\ell = \psi(G)$. Hence, we may assume that $\chi(G) < \ell < \psi(G)$. Suppose that $\psi(G) = k$. Then there is a sequence

$$G = G_0, G_1, \dots, G_t = K_k$$

of graphs where $G_i = \epsilon_i(G_{i-1})$ for some elementary homomorphism ϵ_i of G_{i-1} for $1 \le i \le t$. Since $\chi(G_0) < \ell < \chi(G_t) = k$, there exists a largest integer j with $0 \le j < t$ such that $\chi(G_j) < \ell$. Hence, $\chi(G_{j+1}) \ge \ell$. By Theorem 7.3.2,

$$\chi(G_{j+1}) \le \chi(G_j) + 1 < \ell + 1,$$

and so $\chi(G_{j+1}) = \ell$.   ∎

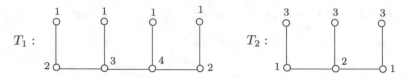

Fig. 7.9 Complete and Grundy colorings

7.4   The Grundy Number of a Graph

A 1939 article [4] by Patrick Michael Grundy dealing with combinatorial games contained ideas that led to the concept of Grundy colorings of graphs. A *Grundy coloring* of a graph G is a proper vertex coloring of G (whose colors, as usual, are positive integers) having the property that for every two colors i and j with $i < j$, every vertex colored j has a neighbor colored i. Consequently, every Grundy coloring is a complete coloring; indeed, it is stronger than a complete coloring. The 4-coloring of the tree T_1 of Fig. 7.9 is a Grundy 4-coloring and is therefore a complete 4-coloring as well. However, the complete 3-coloring of T_2 shown in Fig. 7.9 is not a Grundy 3-coloring.

A *greedy coloring* c of a graph G is obtained from an ordering $\phi : v_1, v_2, \ldots, v_n$ of the vertices of G in some manner, by defining $c(v_1) = 1$, and once colors have been assigned to $v_1, v_2, \ldots, v_t$ for some integer t with $1 \leq t < n$, $c(v_{t+1})$ is defined as the smallest color not assigned to any neighbor of v_{t+1} belonging to the set $\{v_1, v_2, \ldots, v_t\}$. The coloring c so produced is then a Grundy coloring of G. That is, every greedy coloring is a Grundy coloring.

The maximum positive integer k for which a graph G has a Grundy k-coloring is denoted by $\Gamma(G)$ and is called the *Grundy chromatic number* of G or more simply the *Grundy number* of G. If the Grundy number of a graph G is k, then in any Grundy k-coloring of G (using the colors $1, 2, \ldots, k$), every vertex v of G colored k must be adjacent to a vertex colored i for each integer i with $1 \leq i < k$. Thus, $\Delta(G) \geq \deg v \geq k - 1$ and so

$$\Gamma(G) \leq \Delta(G) + 1$$

for every graph G. Since $\Delta(T_1) = 3$ for the tree T_1 in Fig. 7.9, it follows that $\Gamma(T_1) \leq 4$. On the other hand, T_1 has a Grundy 4-coloring and so $\Gamma(T_1) \geq 4$. Therefore, $\Gamma(T_1) = 4$.

Since every Grundy coloring of a graph G is a proper coloring, it follows that

$$\chi(G) \leq \Gamma(G).$$

Christen and Selkow determined those integers k for which a given graph G has a Grundy k-coloring.

Theorem 7.4.1 ([2]) *For a graph G and an integer k with $\chi(G) \le k \le \Gamma(G)$, there is a Grundy k-coloring of G.*

Only a few connected graphs have Grundy number 2.

Theorem 7.4.2 *If G is a connected graph with Grundy number 2, then G is a complete bipartite graph.*

Proof Since G has Grundy number 2 and $\chi(G) \le \Gamma(G)$, it follows that $\chi(G) = 2$ and so G is bipartite. We show that G does not contain P_4 as an induced subgraph. Suppose it does. Let $P = (v_1, v_2, v_3, v_4)$ be an induced subgraph of G, where

$$V(G) - V(P) = \{v_5, v_6, \dots, v_n\}.$$

Consider the sequence

$$\phi : v_1, v_2, v_4, v_3, v_5, v_6, \dots, v_n.$$

The resulting greedy coloring determined by ϕ is a Grundy k-coloring for some $k \ge 3$, which contradicts $\Gamma(G) = 2$. Hence, G does not contain P_4 as an induced subgraph. Thus, G is a complete bipartite graph. ∎

Since a Grundy coloring of a graph G is both a complete coloring and a proper vertex coloring, it follows that

$$\chi(G) \le \Gamma(G) \le \psi(G)$$

for every graph G. Figure 7.10 shows a graph G together with a proper 3-coloring, a Grundy 4-coloring, and a complete 5-coloring of G. Therefore, $\chi(G) \le 3$, $\Gamma(G) \ge 4$, and $\psi(G) \ge 5$. Since G contains an odd cycle, $\chi(G) = 3$; since $\Delta(G) = 3$, $\Gamma(G) = 4$; and since the size of G is $10 < \binom{6}{2}$, $\psi(G) = 5$. The graph G of Fig. 7.10 serves to illustrate a result due to Chartrand, Okamoto, Tuza, and Zhang.

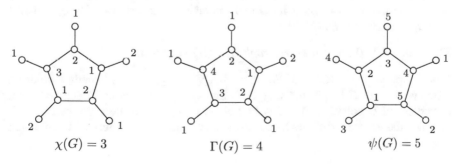

$$\chi(G) = 3 \qquad \Gamma(G) = 4 \qquad \psi(G) = 5$$

Fig. 7.10 Complete and Grundy colorings

Theorem 7.4.3 ([1]) *For integers a, b, c with $2 \le a \le b \le c$, there exists a connected graph G with $\chi(G) = a$, $\Gamma(G) = b$, and $\psi(G) = c$ if and only if $a = b = c = 2$ or $b \ge 3$.*

7.5 The Ochromatic Number of a Graph

In 1982 Simmons [8] introduced a new type of coloring of a graph G based on orderings of the vertices of G, which is similar to but not identical to greedy colorings of G. Let $\phi : v_1, v_2, \ldots, v_n$ be an ordering of the vertices of a graph G. A proper vertex coloring $c : V(G) \to \mathbb{N}$ of G is a *parsimonious ϕ-coloring* of G if the vertices of G are colored in the order ϕ, beginning with $c(v_1) = 1$, such that each vertex v_{i+1} ($1 \le i \le n-1$) must be assigned a color that has been used to color one or more of the vertices $v_1, v_2, \ldots, v_i$ if possible. If v_{i+1} can be assigned more than one color, then a color must be selected that results in using the fewest number of colors needed to color G. If v_{i+1} is adjacent to vertices of every currently used color, then $c(v_{i+1})$ is defined as the smallest positive integer not yet used. The *parsimonious ϕ-coloring number $\chi_\phi(G)$* of G is the minimum number of colors in a parsimonious ϕ-coloring of G. The maximum value of $\chi_\phi(G)$ over all orderings ϕ of the vertices of G is the *ordered chromatic number* or, more simply, the *ochromatic number* of G, which is denoted by $\chi^o(G)$.

To illustrate these concepts, consider the graph $G = P_5$ shown in Fig. 7.11. First, let $\phi_1 : v_1, v_2, v_5, v_3, v_4$. Necessarily, v_1 must be colored 1 and v_2 must be colored 2. Since v_5 is adjacent to neither v_1 nor v_2, it follows that v_5 must be assigned a color already used, that is, v_5 must be colored 1 or 2. If v_5 is colored 2, then v_3 must be colored 1 and v_4 must be colored 3. On the other hand, If v_5 is colored 1, then v_3 must be colored 1 and v_4 must be colored 2. Thus, $\chi_{\phi_1}(G) = 2$. Suppose next that $\phi_2 : v_1, v_4, v_2, v_5, v_3$. Thus, v_1 and v_4 must be colored 1, and v_2 and v_5 must be colored 2. Furthermore, v_3 must be colored 3. Thus, $\chi_{\phi_2}(G) = 3$. There is no ordering ϕ of the vertices of G such that $\chi_\phi(G) = 4$ because $\Delta(G) = 2$ and so no vertex of G will ever be required to be colored 4. Thus, $\chi^o(G) = 3$.

Hedetniemi with Erdös, Hare, and Laskar showed that the ochromatic number of every graph always equals its Grundy number. This fact was also established (unpublished) by Ernest Brickell.

Theorem 7.5.1 ([3]) *For every graph G, $\Gamma(G) = \chi^o(G)$.*

Proof Suppose that $\Gamma(G) = k$. We show that $\chi^o(G) \ge k$. Let a Grundy k-coloring of the vertices of G be given, using the colors $1, 2, \ldots, k$, and let V_i denote the set of vertices of G colored i ($1 \le i \le k$). Let $\phi : v_1, v_2, \ldots, v_n$ be any ordering of G in which the vertices of V_1 are listed first in some order, the vertices of V_2 are listed

Fig. 7.11 Computing the ochromatic number of a graph

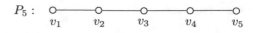

P_5 : v_1 v_2 v_3 v_4 v_5

next in some order, and so on until finally listing the vertices of V_k in some order. We now compute $\chi_\phi(G)$. Assign v_1 the color 1. Since V_1 is independent, every vertex in ϕ that belongs to V_1 is not adjacent to v_1 and must be colored 1 as well. Assume, for an integer r with $1 \leq r < k$, that the parsimonious coloring has assigned the color i to every vertex in V_i for $1 \leq i \leq r$. We now consider the vertices in ϕ that belong to V_{r+1}. Let v_a be the first vertex appearing in ϕ that belongs to V_{r+1}. Since v_a is adjacent to at least one vertex in V_i for every i with $1 \leq i \leq r$, it follows that v_a cannot be colored any of the colors $1, 2, \ldots, r$. Hence, the new color $r + 1$ is assigned to v_a. Now if v_b is any vertex belonging to V_{r+1} such that $b > a$, then v_b cannot be colored any of the colors $1, 2, \ldots, r$ since v_b is adjacent to at least one vertex in V_i for $1 \leq i \leq r$. However since v_b is not adjacent to v_t for all t with $a \leq t < b$, it follows that v_b must be colored $r + 1$. By mathematical induction, $\chi_\phi(G) = k$. Thus, $\chi^o(G) \geq \Gamma(G)$.

We now show that $\Gamma(G) \geq \chi^o(G)$. Let $\phi : v_1, v_2, \ldots, v_n$ be an ordering of the vertices of G such that $\chi_\phi(G) = \chi^o(G)$. Consider the parsimonious ϕ-coloring that is a greedy coloring, that is, whenever there is a choice of a color for a vertex, the smallest possible color is chosen. Suppose that this results in an ℓ-coloring of G. Then $\chi_\phi(G) \leq \ell$. Furthermore, this ℓ-coloring is a Grundy ℓ-coloring. Therefore, $\Gamma(G) \geq \ell$ and so

$$\chi^o(G) = \chi_\phi(G) \leq \ell \leq \Gamma(G),$$

producing the desired inequality.   ∎

Theorem 7.5.1 therefore tells us that the ochromatic number is not a new coloring number but rather an alternative interpretation of the Grundy number.

References

1. G. Chartrand, F. Okamoto, Z. Tuza, P. Zhang, A note on graphs with prescribed complete coloring numbers. J. Comb. Math. Comb. Comput. **73**, 77–84 (2010)
2. C.A. Christen, S.M. Selkow, Some perfect coloring properties of graphs. J. Comb. Theory Ser. B **27**, 49–59 (1979)
3. P. Erdős, W.R. Hare, S.T. Hedetniemi, R. Laskar, On the equality of the Grundy and ochromatic numbers of graphs. J. Graph Theory **11**, 157–159 (1987)
4. P.M. Grundy, Mathematics and games. Eureka **2**, 6–8 (1939)
5. F. Harary, S.T. Hedetniemi, G. Prins, An interpolation theorem for graphical homomorphisms. Port. Math. **26**, 453–462 (1967)
6. S.T. Hedetniemi, Homomorphisms of graphs and automata. Doctoral Dissertation, The University of Michigan, 1966
7. P. Hell, D.J. Miller, Graphs with given achromatic number. Discret. Math. **16**, 195–207 (1976)
8. G.J. Simmons, The ordered chromatic number of planar maps. Congr. Numer. **36**, 59–67 (1982)

Chapter 8
Color Connection and Disconnection

Much of the research in graph theory has dealt with connected graphs. While there are many measures of connectedness for graphs, the two best known are the (vertex) connectivity and edge connectivity, the definitions of which deal with the concepts of vertex-cuts and edge-cuts, respectively. In this chapter, we discuss topics related to these two types of connectedness where coloring also plays a role in each instance.

8.1 Vertex-Cuts and Partition Graphs

There are numerous examples in the graph theory literature where it has been of interest to investigate partitions of the vertex set of a graph (also referred to as partitions of the graph) so that each element (vertex subset) of the partition satisfies some property of interest. For example, when each subset in a partition is an independent set of vertices, we are dealing with a graph homomorphism or proper coloring; while when each subset in a partition induces a connected subgraph, we are dealing with a graph contraction. Each partition π of a graph gives rise to another graph G_π called the *partition graph* (*with respect to* π) of G or the π-*graph* of G where $V(G_\pi) = \pi$, that is, the vertices of G_π are the elements of π. Two distinct vertices V_1 and V_2 are adjacent in G_π if there are adjacent vertices v_1 and v_2 in G where $v_i \in V_i$ for $i = 1, 2$. If $\pi = \{\{v\} : v \in V(G)\}$, then $G_\pi \cong G$; while if $\pi = \{V(G)\}$, then $G_\pi = K_1$. Consequently, if we consider all partitions π of a graph G, then the corresponding collection $\{G_\pi\}$ of graphs includes K_1, the graph G itself, and all graphs that are, in a sense, "between" K_1 and G. In this section, we consider partitions π of a connected graph G where each element of π satisfies a particular connectivity property and the resulting π-graphs of G.

A set U of vertices in a connected graph G is a *vertex-cut* if $G - U$ is disconnected. Provided G is not complete, G has at least one vertex-cut. If U is a vertex-cut of G with a minimum number of vertices, then U is a *minimum vertex-cut* and $|U|$ is the *connectivity* $\kappa(G)$ of G. If no proper subset of U is a vertex-cut of G, then U is a *minimal vertex-cut.* If there exists a partition $\pi = \{V_1, V_2, \ldots, V_k\}$ of G such that each set V_i is a minimal vertex-cut of G, then π is called an $\mathcal{MVC}$-*partition* of G.

As an illustration of graphs possessing $\mathcal{MVC}$-partitions, we consider two classes of graphs, namely cycles and prisms, the latter of which are Cartesian products of a cycle with K_2. The 8-cycle $F = C_8$, shown in Fig. 8.1, has several $\mathcal{MVC}$-partitions, each consisting of four pairs of nonadjacent vertices. For example, the set $\pi_1 = \{V_1, V_2, V_3, V_4\}$, where

$$V_1 = \{w_1, w_4\}, V_2 = \{w_2, w_8\}, V_3 = \{w_3, w_6\}, \text{ and } V_4 = \{w_5, w_7\},$$

is an $\mathcal{MVC}$-partition. The partition graph $F_{\pi_1} = K_4$ is also shown in Fig. 8.1. In fact, every even cycle has an $\mathcal{MVC}$-partition while no odd cycle has one.

The prism $H = C_9 \,\square\, K_2$, shown in Fig. 8.2, is a 3-regular graph having connectivity 3. Therefore, for each vertex v of H, the neighborhood $N(v)$ of v is a minimal vertex-cut of H. Each of the five sets

$$U_1 = N(u_1) = \{v_1, u_2, u_9\}, U_2 = N(v_1) = \{u_1, v_2, v_9\},$$
$$U_3 = \{u_3, v_3, u_8, v_8\}, U_4 = \{u_4, v_4, u_6, v_6\}, \text{ and } U_5 = \{u_5, v_5, u_7, v_7\}$$

is a minimal vertex-cut of H. Therefore, $\pi_2 = \{U_1, U_2, U_3, U_4, U_5\}$ is an $\mathcal{MVC}$-partition of H. The partition graph H_{π_2} is also shown in Fig. 8.2.

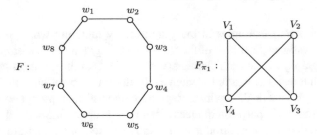

Fig. 8.1 The graph F and a partition graph of F

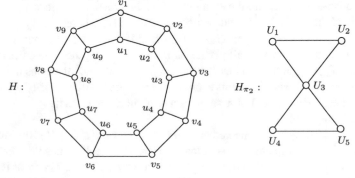

Fig. 8.2 A graph H and a partition graph of H

If π is an $\mathcal{MVC}$-partition of G, then the π-graph G_π provides information on how the minimal vertex-cuts in π are related to one another. In particular, for every connected graph G with an $\mathcal{MVC}$-partition π, the graph G_π is also connected. The concept of the π-graph of a graph with an $\mathcal{MVC}$-partition π was introduced by Hedetniemi along with Chartrand, Haynes, and Zhang in [2].

The following theorem characterizes all graphs that are π-graphs of even cycles for an $\mathcal{MVC}$-partition π. For two disjoint sets U and W of vertices in a graph G, the set of edges joining U and W in G is denoted by $[U, W]$. The *underlying graph* of a multigraph M is the graph G for which $V(G) = V(M)$ and $uv \in E(G)$ if u and v are joined by at least one edge in M.

Theorem 8.1.1 ([2]) *A connected graph H of order 3 or more is a π-graph of some even cycle G with an $\mathcal{MVC}$-partition π if and only if H is the underlying graph of a 4-regular multigraph.*

Proof First, assume that H is the underlying graph of a 4-regular multigraph M of order $k \geq 3$. Thus, M is an Eulerian multigraph of even size $2k$. Let $C = (v_1, v_2, \ldots, v_{2k}, v_1)$ be an Eulerian circuit in M. Since M is 4-regular, each vertex of M occurs exactly twice as nonconsecutive vertices of C. Next, let $G = C_{2k} = (u_1, u_2, \ldots, u_{2k}, u_1)$ be a $2k$-cycle. Furthermore, let $\pi = \{V_1, V_2, \ldots, V_k\}$ be the $\mathcal{MVC}$-partition of G in which two vertices u_a and u_b of G belong to the same element of π if and only if $v_a = v_b$ on the circuit C. Then two vertices V_i and V_j of G_π are adjacent if and only if some vertex in V_i is adjacent to some vertex of V_j, where $1 \leq i < j \leq k$. Let $V_i = \{u_a, u_b\}$ and $V_j = \{u_c, u_d\}$. Now, V_i and V_j are adjacent if and only if $\|[V_i, V_j]\| > 1$. Thus, $H \cong G_\pi$.

For the converse, assume that H is a connected graph of order k that is a π-graph of some even cycle G for an $\mathcal{MVC}$-partition π of G. Necessarily, G has order $2k$ and so $G = (u_1, u_2, \ldots, u_{2k}, u_1)$ is a $2k$-cycle. Let $\pi = \{V_1, V_2, \ldots, V_{2k}\}$ be an $\mathcal{MVC}$-partition of $V(G)$ such that $G_\pi \cong H$. For each integer i with $1 \leq i \leq k$, let $V_i = \{u_{i_1}, u_{i_2}\}$ where then u_{i_1} and u_{i_2} are two nonadjacent vertices of G. Let M be the multigraph with $V(M) = \pi$, where the number of edges joining V_i and V_j is $\|[V_i, V_j]\|$ for $i \neq j$. Thus, H is the underlying graph of M. Since the two vertices in V_i are nonadjacent, it follows that $\|[V_i, V(G) - V_i]\| = 4$ for $1 \leq i \leq k$ and so M is 4-regular. ∎

The following result is then an immediate consequence of Theorem 8.1.1.

Corollary 8.1.2 ([2])   *Every connected 4-regular graph is a π-graph of an even cycle with an $\mathcal{MVC}$-partition π.*

With the aid of Theorem 8.1.1, those cubic graphs that are π-graphs of an even cycle with an $\mathcal{MVC}$-partition π can be determined. A *1-factor* of a graph G is a 1-regular spanning subgraph of G.

Corollary 8.1.3 ([2])   *A connected cubic graph H is a π-graph of an even cycle with an $\mathcal{MVC}$-partition π if and only if H has a 1-factor.*

Proof Suppose that H is a connected cubic graph having a 1-factor F. By replacing each edge in F by two parallel edges, a 4-regular multigraph M is obtained. Thus, H is the underlying graph of M. It then follows by Theorem 8.1.1 that H is a π-graph of an even cycle with an $\mathcal{MVC}$-partition π.

For the converse, assume that H is a connected cubic graph that is a π-graph of an even cycle G for an $\mathcal{MVC}$-partition π of G. Then H has even order, say $2k \geq 4$. By Theorem 8.1.1, H is the underlying graph of a 4-regular multigraph M. Let v_1 be a vertex of H. Since $\deg_H v_1 = 3$ and $\deg_M v_1 = 4$, it follows that v_1 is incident with exactly one edge e_1 in M that does not belong to H, say $e_1 = v_1 w_1$. Hence, e_1 is the only edge of M that is incident with w_1 that does not belong to H. Continuing in this manner, we obtain k pairwise nonadjacent edges $e_i = u_i w_i$ $(1 \leq i \leq k)$ that belong to M but not to H. Therefore, the edges $u_i w_i$ $(1 \leq i \leq k)$ of H form a 1-factor of H. ∎

For example, the famous Petersen graph P is a connected cubic graph of order 10 containing a 1-factor. Therefore, by Corollary 8.1.3, P is the π-graph of the cycle C_{20} for some $\mathcal{MVC}$-partition π. This is illustrated in Fig. 8.3, where each edge of a 1-factor F of P is replaced by two parallel edges, resulting in a 4-regular multi-graph M. An Eulerian circuit C in M is

$$C = (v_1, v_6, v_8, v_{10}, v_7, v_9, v_6, v_1, v_2, v_7, v_2, v_3, v_8, v_3, v_4, v_9, v_4, v_5, v_{10}, v_5, v_1).$$

Let $C_{20} = (u_1, u_2, \ldots, u_{20}, u_1)$ be a cycle of order 20. Following the proof of Theorem 8.1.1, an $\mathcal{MVC}$-partition $\pi = \{V_1, V_2, \ldots, V_{10}\}$ of C_{20} is constructed, where

$$V_1 = \{u_1, u_8\},\ V_2 = \{u_9, u_{11}\},\ V_3 = \{u_{12}, u_{14}\},$$
$$V_4 = \{u_{15}, u_{17}\},\ V_5 = \{u_{18}, u_{20}\},\ V_6 = \{u_2, u_7\},$$
$$V_7 = \{u_5, u_{10}\},\ V_8 = \{u_3, u_{13}\},\ V_9 = \{u_6, u_{16}\},\ V_{10} = \{u_4, u_{19}\}.$$

The Petersen graph is the π-graph G_π of $G = C_{20}$, shown in Fig. 8.3.

Fundamental questions arise from this discussion of $\mathcal{MVC}$-partitions of graphs.

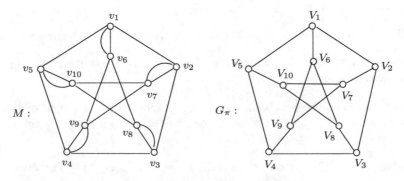

Fig. 8.3 Showing that the Petersen graph is a π-graph of an even cycle

Problem 8.1.4 Which connected graphs have $\mathcal{MVC}$-partitions?

Problem 8.1.5 Let G be a connected graph with $\mathcal{MVC}$-partitions. Among all such partitions π of G, what are the minimum and maximum values of $|\pi|$?

We saw that if π is an $\mathcal{MVC}$-partition of an even cycle, then each element of π is not only a minimal vertex-cut, it is an independent set. In fact, if π is a partition of the vertex set of a graph G such that each element of π is an independent set, then both π and G_π concern familiar concepts in graph theory.

Recall (from Chap. 7) that a *homomorphism* from a graph G to a graph H' is a function $\phi : V(G) \to V(H')$ that maps adjacent vertices in G to adjacent vertices in H. Also, the subgraph $H = (V, E)$ of H' whose vertex set is $V(H) = \phi(V(G))$ and whose edge set is $E(H) = \{\phi(u)\phi(v) : uv \in E(G)\}$ is the *homomorphic image* of G under ϕ and is denoted by $\phi(G) = H$. A graph H is called a *homomorphic image* of a graph G if there is a homomorphism ϕ of G such that $\phi(G) = H$.

Let $H = (V, E)$ be a homomorphic image of a graph G and let $V(H) = \{v_1, v_2, \ldots, v_k\}$. For any vertex $v \in V(H)$, let $\phi^{-1}(v) = \{u \in V(G) : \phi(u) = v\}$. Since ϕ is a homomorphism, it follows that $\phi^{-1}(v)$ is an independent set in G for every $v \in V(H)$. This defines a *coloring* of the vertices of G, that is,

$$\pi = \{\phi^{-1}(v_1), \phi^{-1}(v_2), \ldots, \phi^{-1}(v_k)\}$$

is a partition of G into k independent sets. It also follows that the π-graph G_π is isomorphic to H. In particular, if π is an $\mathcal{MVC}$-partition of a graph G, where each set V_i $(1 \le i \le k)$ is independent, then ϕ represents a k-coloring of G where each vertex of V_i $(1 \le i \le k)$ is colored i and G_π is the homomorphic image of G resulting from this k-coloring.

For a nontrivial connected graph H that is a homomorphic image of a cycle, let $\mu(H)$ be the length of a shortest such cycle and let $e(H)$ be the Eulerian number of H (the length of an Eulerian walk in H), which is a closed walk of minimum length traversing every edge of H. (Eulerian walks are discussed in Chap. 6.)

While, by Theorem 8.1.1, a connected graph H of order 3 or more is a π-graph of a cycle for an $\mathcal{MVC}$-partition π of the vertex set of H only if H is the underlying graph of a 4-regular multigraph, every such graph H is a π-graph of a cycle for some partition π of the vertex set of a cycle into independent sets.

Theorem 8.1.6 ([2]) *Every nontrivial connected graph H is a homomorphic image of a cycle and so $\mu(H)$ exists. Furthermore, $\mu(H) = e(H)$.*

Proof Let H be a nontrivial connected graph with $V(H) = \{u_1, u_2, \ldots, u_k\}$ and let

$$W = (w_1, w_2, \ldots, w_p, w_1)$$

be an Eulerian walk of length $e(H) = p$ in H. Thus, each edge in H occurs in W at least once and at most twice. Let $C = (v_1, v_2, \ldots, v_p, v_1)$ be a cycle of order p. For each integer i with $1 \le i \le k$, let

$$V_i = \{v_t \in V(C) : \ w_t = u_i \text{ and } 1 \leq t \leq p\}.$$

Therefore, each set V_i is an independent set of vertices of H and a vertex in a set V_i is adjacent to a vertex in V_j in C ($1 \leq i \leq j$ and $i \neq j$) if and only if $u_i u_j \in E(H)$. Hence, H is a homomorphic image of C. Therefore, $\mu(H)$ exists and $\mu(H) \leq e(H)$.

Next, we show that $e(H) \leq \mu(H)$. Let $\mu(H) = \ell$ and let

$$C = (v_1, v_2, \ldots, v_\ell, v_1)$$

be a cycle such that H is a homomorphic image of C. Thus, there is a partition $\mathcal{P} = \{U_1, U_2, \ldots, U_k\}$ of $V(C)$ into independent sets such that $u_i u_j \in E(H)$ if and only if some vertex in U_i is adjacent to some vertex in U_j in C. Since each U_i is an independent set of vertices of C, it follows that every edge in C gives rise to an edge in H (although it is possible that several edges in C produce the same edge in H). Furthermore, for each edge $u_i u_j$ in H, there is at least one edge $v_s v_{s+1}$ in C such that $v_s \in U_i$ and $v_{s+1} \in U_j$. Identifying all vertices in each set U_i, producing a single vertex denoted by u_i for $1 \leq i \leq k$, and following the ordering of vertices in C, we obtain an Eulerian walk $W = (w_1, w_2, \ldots, w_\ell, w_1)$ of length ℓ in H. Hence, $e(H) \leq \mu(H)$ and so $e(H) = \mu(H)$.   ∎

We have seen several results dealing with π-graphs of even cycles where π is an $\mathcal{MVC}$-partition and have discussed some π-graphs of the prisms $C_n \ \square \ K_2$. This suggests the following problem.

Problem 8.1.7   Which graphs are π-graphs of a prism $C_n \ \square \ K_2$, where π is an $\mathcal{MVC}$-partition? Investigate homographic images of these prisms.

By Theorem 8.1.6, every nontrivial connected graph is a homomorphic image of a connected 2-regular graph. The following result was obtained by Chartrand, Haynes, Hedetniemi, and Zhang.

Theorem 8.1.8   ([2])   *For every pair k, n of positive integers where $n \geq 3$, the graph $C_n \ \square \ Q_k$ is a homomorphic image of $C_n \ \square \ Q_{k+1}$.*

The following is then a consequence of Theorems 8.1.6 and 8.1.8.

Corollary 8.1.9   ([2])   *Every nontrivial connected graph is a homomorphic image of an r-regular graph for each integer $r \geq 2$.*

Theorem 8.1.8 also states that H is a homomorphic image of $H \ \square \ K_2$ for $H = C_n \ \square \ Q_k$. This suggests the problem of determining those nontrivial connected graphs H having the property that H is a homomorphic image of $H \ \square \ K_2$. For a vertex v in a connected graph G, recall that $e(v)$ denotes the *eccentricity* of v (the largest distance from v to a vertex in G), which was described in Chap. 5.

Theorem 8.1.10   ([2])   *Every nontrivial tree T is a homomorphic image of $T \ \square \ K_2$.*

Proof Let T be a tree of order $n \geq 2$ and let v_1 be a leaf of T. The tree T may then be considered as a rooted tree with root v_1. Therefore, T can be considered as a directed tree where there is a directed $v_1 - w$ path in T for every vertex w of T. Let $V(T) = \{v_1, v_2, \ldots, v_n\}$ where $d(v_1, v_i) \leq d(v_1, v_j)$ for $1 \leq i \leq j \leq n$. Let $G = T \,\square\, K_2$, where G consists of the tree T (as labeled above), a second copy T' of T with $V(T') = \{u_1, u_2, \ldots, u_n\}$ such that u_i corresponds to v_i and $u_i v_i \in E(G)$ for $1 \leq i \leq n$.

We now show that there is a proper n-coloring c of G using the colors $1, 2, \ldots, n$ resulting in the color classes $V_1, V_2, \ldots, V_n$ such that the homomorphic image resulting from the n color classes $V_1, V_2, \ldots, V_n$ is isomorphic to T. First, color each vertex v_i the color i for $i = 1, 2, \ldots, n$.

The vertex u_2 is the only vertex at distance 1 from u_1 of T'. Assign the color 2 to u_1 and the color 1 to u_2. Next, assign the color 2 to each vertex of T' at distance 1 from u_2. Proceeding recursively, assume that all vertices of T' at distance k from u_1 have been assigned a color where $2 \leq k < e_{T'}(u_1)$ and let $u_j \in V(T')$ such that $d_{T'}(u_1, u_j) = k + 1$. Let u_i be the unique vertex adjacent to u_j on the $u_1 - u_j$ (directed) path P on T'. We then assign the color $c(v_i)$ to u_j (and so $c(u_j) = c(v_i)$).

Let a and b be distinct colors in $\{1, 2, \ldots, n\}$ such that some vertex in V_a is adjacent to a vertex in V_b. Then $v_a v_b \subset E(T)$ where, say, $a < b$. Thus, the edge u_b is colored a. Also, the two incident vertices of an edge of T' are assigned two distinct colors of $\{1, 2, \ldots, n\}$ if and only if the two incident vertices of some edge of T are also assigned these same two colors. Hence, T is a homomorphic image of G.   ∎

Although every tree T is a homomorphic image of $T \,\square\, K_2$, not every nontrivial connected graph G is a homomorphic image of $G \,\square\, K_2$. For example, it was shown in [2] that the graph H of Fig. 8.4 is not a homomorphic image of $H \,\square\, K_2$.

These concepts and results suggest many problems, particularly those dealing with the structure of π-graphs obtained from partitions $\pi = \{V_1, V_2, \ldots, V_k\}$ of the vertex set of a connected graph where the elements of π have some property of interest. If each set V_i is an independent set, then these partitions give rise to proper colorings of graphs and the corresponding π-graphs are homomorphic images of a graph (as described in Chap. 7). If each set V_i induces a connected subgraph (a connected partition), then the corresponding π-graphs are *contractions* of a graph. In this case, a famous conjecture of Hadwiger comes to mind.

Fig. 8.4 The graph H

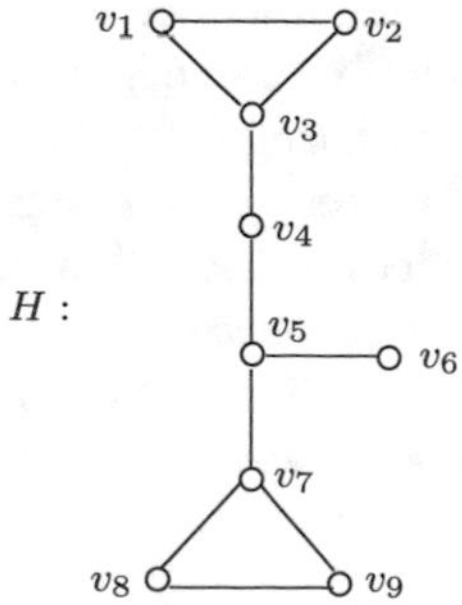

H :

Hadwiger's Conjecture ([6]) *If G is a connected graph with chromatic number k, then G has a connected partition π whose corresponding π-graph is the complete graph K_k.*

While Hadwiger's conjecture is known to be true for $1 \leq k \leq 6$, this is not known when $k \geq 7$.

A more general problem is suggested by the concepts described in this section.

Problem 8.1.11   For graphs H and G, determine whether $G_\pi \cong H$ for some partition π of G.

8.2   Edge-Cuts and Rainbow Disconnection

While a nontrivial connected graph contains a vertex-cut only if it is not complete, every nontrivial connected graph contains edge-cuts. An *edge-cut* of a nontrivial connected graph G is a set R of edges of G such that $G - R$ is disconnected. The minimum number of edges in an edge-cut of G is its *edge-connectivity* $\lambda(G)$. We then have the well-known inequality $\lambda(G) \leq \delta(G)$, where $\delta(G)$ is the minimum degree of G. For two distinct vertices u and v of G, let $\lambda(u, v)$ denote the minimum number of edges in an edge-cut R of G such that u and v lie in different components of $G - R$. The following result of Elias, Feinstein, and Shannon [4] and Ford and Fulkerson [5] presents a different interpretation of $\lambda(u, v)$.

Theorem 8.2.1   ([4, 5]) *For every two vertices u and v in a graph G, $\lambda(u, v)$ is the maximum number of pairwise edge-disjoint $u - v$ paths in G.*

The *upper edge-connectivity* $\lambda^+(G)$ is defined by

$$\lambda^+(G) = \max\{\lambda(u, v) :\ u, v \in V(G)\}.$$

Consider, for example, the graph $K_n + v$ obtained from the complete graph K_n, one vertex of which is attached to a single leaf v. For this graph, $\lambda(K_n + v) = 1$ while $\lambda^+(K_n + v) = n - 1$. Thus, $\lambda(G)$ denotes the global minimum edge-connectivity of a graph, while $\lambda^+(G)$ denotes the local maximum edge-connectivity of a graph.

A nonempty graph G is *edge-colored* if every edge of G is assigned a color from some prescribed set of colors. A set X of edges in an edge-colored graph is a *rainbow set* if no two edges of X have the same color. In 2006, Chartrand, Johns, McKeon, and Zhang defined an edge-colored connected graph G to be *rainbow-connected* if for every two vertices u and v of G, there exists a rainbow $u - v$ path in G (all edges of the path have different colors). The minimum number of colors required for a connected graph G to be rainbow-connected is the *rainbow connection number* of G. The first paper [3] on this topic was published in 2008. Since then, the concept has been studied extensively and, in fact, a book by Li and Sun [7] has been written on this topic.

A set R of edges in a nontrivial connected edge-colored graph G is a *rainbow cut* of G if R is both a rainbow set and an edge-cut. A rainbow cut R is said to *separate* two vertices u and v of G if u and v belong to different components of $G - R$. Any such rainbow cut in G is called a $u - v$ *rainbow cut* in G. An edge-coloring of G is a *rainbow disconnection coloring* if for every two distinct vertices u and v of G, there exists a $u - v$ rainbow cut in G. The *rainbow disconnection number* rd(G) of G is the minimum number of colors required of a rainbow disconnection coloring of G. Consequently, if G is a connected graph of order $n \geq 2$, then $1 \leq$ rd$(G) \leq n - 1$. A rainbow disconnection coloring with rd(G) colors is called an rd-*coloring* of G. These concepts were introduced by Hedetniemi along with Chartrand, Devereaux, Haynes, and Zhang in [1]. The following result gives bounds for the rainbow disconnection number of a graph. Once again, recall that $\chi'(G)$ denotes the chromatic index of G and $\Delta(G)$ the maximum degree of G.

Proposition 8.2.2 ([1]) *If G is a nontrivial connected graph, then*

$$\lambda(G) \leq \lambda^+(G) \leq \text{rd}(G) \leq \chi'(G) \leq \Delta(G) + 1.$$

As an illustration of Proposition 8.2.2, we show that every cycle C_n of order $n \geq 3$ has rainbow disconnection number 2. Since rd$(C_n) \geq \lambda(C_n) = 2$ by Proposition 8.2.2, it suffices to show that C_n has a rainbow disconnection coloring using two colors. Let c be an edge-coloring of C_n that assigns the color 1 to exactly $n - 1$ edges of C_n and the color 2 to the remaining edge e of C_n. Let u and v be two vertices of C_n. There are two $u - v$ paths P and Q in C_n, exactly one of which contains the edge e, say $e \in E(P)$. Then any set $\{e, f\}$, where $f \in E(Q)$, is a $u - v$ rainbow cut. Thus, c is a rainbow disconnection coloring of C_n. Hence, rd$(C_n) = 2$. It was shown in [1] that rd$(W_n) = 3$, where W_n is the wheel of order $n + 1 \leq 4$ (the join of C_n and K_1). Figure 8.5 shows rainbow disconnection colorings of the wheels W_6 and W_7 using three colors. In each case, the set E_{v_i} of the three edges incident with v_i is a rainbow set for $1 \leq i \leq n$ where $n = 6, 7$. Furthermore, for every two distinct vertices x and y of W_n, at least one of x and y belongs to C_n, say $x \in V(C_n)$, and then E_x separates x and y and so E_x is an $x - y$ rainbow cut.

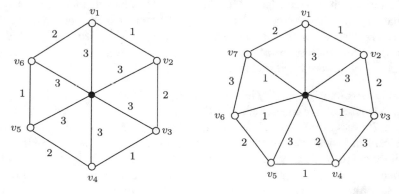

Fig. 8.5 Rainbow disconnection colorings of W_6 and W_7

Since $\chi'(C_n) = 3$ when n is odd and $\chi'(W_n) = n$ for each integer $n \geq 3$, it follows that $\mathrm{rd}(G) < \chi'(G)$ if G is an odd cycle or if G is a wheel of order at least 4. Wheels therefore show that there are graphs G for which $\chi'(G) - \mathrm{rd}(G)$ can be arbitrarily large. There are also graphs G for which $\lambda^+(G) < \mathrm{rd}(G) = \chi'(G)$.

Proposition 8.2.3 ([1]) *The rainbow disconnection number of the Petersen graph is* 4.

Proof Let P denote the Petersen graph where $V(P) = \{v_1, v_2, \ldots, v_{10}\}$. Since $\lambda(P) = 3$ and $\chi'(P) = 4$, it follows by Proposition 8.2.2 that either $\mathrm{rd}(P) = 3$ or $\mathrm{rd}(P) = 4$. Assume, to the contrary, that $\mathrm{rd}(P) = 3$ and let there be given a rainbow disconnection 3-coloring of P. Now, let u and v be two vertices of P and let R be a $u - v$ rainbow cut. Hence, $|R| \leq 3$ and $P - R$ is disconnected, where u and v belong to different components of $P - R$. Let U be the vertex set of the component of $P - R$ containing u, where $|U| = k$. We may assume that $1 \leq k \leq 5$. First, suppose that $1 \leq k \leq 4$. Since the girth of P is 5, the subgraph $P[U]$ induced by U contains $k - 1$ edges. Therefore, $|R| = 3k - (2k - 2) = k + 2$, where then $3 \leq k + 2 \leq 6$. If $k = 5$, then $P[U]$ contains at most five edges and so $|R| \geq 5$, which is impossible. Since $\mathrm{rd}(P) = 3$, it follows that $|R| \leq 3$ and so $k = 1$. Hence, the only possible $u - v$ rainbow cut is the set of the three edges incident with u (or with v).

Let the colors assigned to the edges of P be red, blue, and green. Since $\chi'(P) = 4$, there is at least one vertex of P that is incident with two edges of the same color. We claim, in fact, that there are at least two such vertices. Let E_R, E_B, and E_G denote the sets of edges of P colored red, blue, and green, respectively, and let P_R, P_B, and P_G be the spanning subgraphs of P with edge sets E_R, E_B, and E_G. We may assume that

$$|E_R| \geq |E_B| \geq |E_G| \text{ and so } |E_R| \geq 5.$$

* If $|E_R| \geq 7$, then $\sum_{i=1}^{10} \deg_{P_R} v_i \geq 14$. Since $\deg_{P_R} v_i \leq 3$ for each i with $1 \leq i \leq 10$, at least two vertices are incident with two red edges, verifying the claim.
* If $|E_R| = 6$, then $\sum_{i=1}^{10} \deg_{P_R} v_i = 12$. Then either (i) at least two vertices are incident with two red edges or (ii) there is a vertex, say v_{10}, incident with three red edges and each of $v_1, v_2, \ldots, v_9$ is incident with exactly one red edge. If (ii) occurs, then either $|E_B| = 6$ or $|E_B| = 5$ and so $\sum_{i=1}^{9} \deg_{P_B} v_i \geq 10$, which implies that at least one of the vertices $v_1, v_2, \ldots, v_9$ is incident with two blue edges, again verifying the claim.

The only remaining possibility is therefore $|E_R| = |E_B| = |E_G| = 5$. If E_R is an independent set of five edges, then $P - E_R$ is a 2-regular graph. Since the girth of P is 5 and P is not Hamiltonian, it follows that $P - E_R$ consists of two vertex-disjoint 5-cycles. Thus, there is a vertex of P in each cycle incident with two blue edges or with two green edges, verifying the claim. Hence, none of E_R, E_B, or E_G is an independent set. This implies that for each of these colors, there is a vertex of P incident with two edges of this color, verifying the claim in general.

Thus, P contains two vertices u and v, each of which is incident with two edges of the same color. Since the only $u - v$ rainbow cut is the set of edges incident with u or v, this is a contradiction.   ∎

The following two results describe those graphs having rainbow disconnection number 1 or 2 and those graphs of order n having rainbow disconnection number $n-1$.

Theorem 8.2.4 ([1]) *Let G be a nontrivial connected graph. Then*

(1) $\mathrm{rd}(G) = 1$ *if and only if G is a tree and*
(2) $\mathrm{rd}(G) = 2$ *if and only if each block of G is either K_2 or a cycle, and at least one block of G is a cycle.*

Theorem 8.2.5 ([1])   *Let G be a nontrivial connected graph of order n. Then $\mathrm{rd}(G) = n - 1$ if and only if G contains at least two vertices of degree $n - 1$.*

Proof First, suppose that G is a nontrivial connected graph of order n containing at least two vertices of degree $n - 1$. Since $\mathrm{rd}(G) \leq n - 1$, it remains to show that $\mathrm{rd}(G) \geq n - 1$. Let $u, v \in V(G)$ such that $\deg u = \deg v = n - 1$. Among all sets of edges that separate u and v in G, let S be one of minimum size. We show that $|S| \geq n - 1$. Let U be a component of $G - S$ that contains u and let $W = V(G) \setminus U$. Thus, $v \in W$ and $S = [U, W]$ consists of those edges in $G - S$ joining a vertex of U and a vertex of W. Suppose that $|U| = k$ for some integer k with $1 \leq k \leq n - 1$ and then $|W| = n - k$. The vertex u is adjacent to each of the $n - k$ vertices of W and each of the remaining $k - 1$ vertices in U is adjacent to at least one vertex in W. Hence, $|S| \geq n - k + (k - 1) = n - 1$. This implies that every $u - v$ rainbow cut contains at least $n - 1$ edges of G and so $\mathrm{rd}(G) \geq n - 1$.

For the converse, suppose that G is a nontrivial connected graph of order n having at most one vertex of degree $n - 1$. We show that $\mathrm{rd}(G) \leq n - 2$. We consider two cases.

Case 1. Exactly one vertex v of G has degree $n - 1$. Let $H = G - v$. Thus, $\Delta(H) \leq n - 3$. Since $\chi'(H) \leq \Delta(H) + 1 = n - 2$, there is a proper edge-coloring of H using $n - 2$ colors. We now define an edge-coloring $c : E(G) \rightarrow [n - 2]$ of G. First, let c be a proper $(n - 2)$-edge-coloring of H. For each vertex $x \in V(H)$, since $\deg_H x \leq n - 3$, there is $a_x \in [n - 2]$ such that a_x is not assigned to any edge incident with x. Define

$$c(vx) = a_x \text{ for each edge } vx \text{ incident with } x.$$

Thus, the set E_x of edges incident with x is a rainbow set for each $x \in V(H)$. Let u and w be two distinct vertices of G. Then at least one of u and w belongs to H, say $u \in V(H)$. Since E_u separates u and w, it follows that c is a rainbow disconnection coloring of G using $n - 2$ colors. Hence, $\mathrm{rd}(G) \leq n - 2$.

Case 2. No vertex of G has degree $n - 1$. Therefore, $\Delta(G) \leq n - 2$. If $\Delta(G) \leq n - 3$, then $\mathrm{rd}(G) \leq \chi'(G) \leq n - 2$ by Proposition 8.2.2. Thus, we may assume that $\Delta(G) = n - 2$. Suppose first that G is not $(n - 2)$-regular. We claim that G is a connected spanning subgraph of some graph G^* of order n having exactly one vertex of degree $n - 1$. Let u be a vertex of degree $k \le n - 3$ in G. Let $N(u)$ be the neighborhood of u and $W = V(G) \setminus N[u]$, where $N[u] = N(u) \cup \{u\}$ is the closed neighborhood of u. Then $|N(u)| = k$ and $|W| = n - k - 1 \ge 2$. If W contains a vertex v of degree $n - 2$ in G, then v is the only vertex of degree $n - 1$ in $G^* = G + uv$. If no vertex in W has degree $n - 2$ in G, then let G^* be the graph obtained from G by joining u to each vertex in W. In this case, u is the only vertex of degree $n - 1$ in G^*. It then follows by Case 1 that $\mathrm{rd}(G^*) \le n - 2$. Since G is a connected spanning subgraph of G^*, it follows that $\mathrm{rd}(G) \le \mathrm{rd}(G^*) \le n - 2$. Finally, suppose that G is $(n - 2)$-regular. Thus, G is 1-factorable and so $\chi'(G) = \Delta(G) = n - 2$. Therefore, $\mathrm{rd}(G) \le \chi'(G) = n - 2$ by Proposition 8.2.2.   ∎

A problem of interest concerns the extremal values of the size of a graph of a fixed order with prescribed rainbow disconnection number.

Problem 8.2.6 ([1]) For a given pair k, n of positive integers with $k \le n - 1$, what are the minimum possible size and maximum possible size of a connected graph G of order n such that the rainbow disconnection number of G is k?

By Theorem 8.2.4, the only connected graphs of order n having rainbow disconnection number 1 are the trees of order n. That is, the connected graphs of order n having rainbow disconnection number 1 have size $n - 1$. Also, by Theorem 8.2.4, the minimum size of a connected graph of order $n \ge 3$ having rainbow disconnection number 2 is n. Furthermore, by Theorem 8.2.5, the minimum size of a connected graph of order $n \ge 2$ having rainbow disconnection number $n - 1$ is $2n - 3$. These facts are special cases of a more general result. To show this, the following lemma is useful.

Lemma 8.2.7 ([1]) *If x and y are two nonadjacent vertices of a connected graph H that is not complete, then $\mathrm{rd}(H + xy) \le \mathrm{rd}(H) + 1$.*

Proof Suppose that $\mathrm{rd}(H) = k$ and let c_0 be an rd-coloring of H using the colors $1, 2, \ldots, k$. The coloring c_0 is extended to an edge-coloring c of $H + xy$ by assigning the color $k + 1$ to the edge xy. Now, let u and v be two vertices of H and let R be a $u - v$ rainbow cut in H. Then $R \cup \{xy\}$ is a $u - v$ rainbow cut in $H + xy$. Hence, c is a rainbow disconnection $(k + 1)$-coloring of $H + xy$. Therefore, $\mathrm{rd}(H + xy) \le k + 1 = \mathrm{rd}(H) + 1$.   ∎

Theorem 8.2.8 ([1]) *The minimum size of a connected graph of order n having rainbow disconnection number k, where $1 \le k \le n$, is $n + k - 2$.*

Proof Since $\mathrm{rd}(K_n) = n - 1$, the result is true for $k = n - 1$. Hence, we may assume that $1 \le k \le n - 2$. First, we show that if the size of a connected graph G of order n is $n + k - 2$, then $\mathrm{rd}(G) \le k$. We proceed by induction on k. We have seen that the result is true for $k = 1, 2$ by Theorem 8.2.4. Suppose that if the size of a connected graph H of order n is $n + k - 2$ for some integer k with $2 \le k \le n - 3$, then $\mathrm{rd}(H) \le k$. Let G be a connected graph of order n and size $n + (k + 1) - 2 = n + k - 1$. We show that $\mathrm{rd}(G) \le k + 1$. Since G is not a tree, there is an edge e in G such that $H = G - e$

Fig. 8.6 A graph G of order
7 and $\mathrm{rd}(G) = 3$

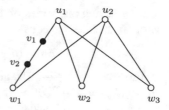

is a connected spanning subgraph of G. Since the size of H is $n + k - 2$, it follows
by the induction hypothesis that $\mathrm{rd}(H) \leq k$. Hence, $\mathrm{rd}(G) = \mathrm{rd}(H + e) \leq k + 1$
by Lemma 8.2.7. Therefore, the minimum possible size of a connected graph G of
order n having $\mathrm{rd}(G) = k$ is $n + k - 2$.

It remains to show that for each pair k, n of integers with $1 \leq k \leq n - 1$ there
is a connected graph G of order n and size $n + k - 2$ such that $\mathrm{rd}(G) = k$. Since
this is true for $k = 1, 2, n - 1$, we now assume that $3 \leq k \leq n - 2$. Let $H = K_{2,k}$
with partite set $U = \{u_1, u_2\}$ and $W = \{w_1, w_2, \ldots, w_k\}$. Now, let G be the graph
of order n and size $n + k - 2$ obtained from H by subdividing the edge $u_1 w_1$ a total
of $n - k - 2$ times, producing the path $(u_1, v_1, v_2, \ldots, v_{n-k-2}, w_1)$ in G. This graph
is illustrated in Fig. 8.6 for $k = 3$ and $n = 7$.

Since $\chi'(H) = k$, there is a proper edge-coloring c_H of H using the colors
$1, 2, \ldots, k$. We may assume that $c(u_1 w_1) = 1$ and $c(u_2 w_1) = 2$. We now extend
the coloring c_H to a proper edge-coloring c_G of G using the colors $1, 2, \ldots, k$ by
defining $c_G(u_1 v_1) = 1$ and alternating the colors of the edges of P with 3 and 1
thereafter. Hence, $\chi'(G) = k$ and so $\mathrm{rd}(G) \leq \chi'(G) = k$ by Proposition 8.2.2.
Furthermore, since $\lambda(u_1, u_2) = k$ and $\lambda(x, y) = 2$ for all other pairs x, y of vertices
of G, it follows that $\lambda^+(G) = k$. Again, by Proposition 8.2.2, $\mathrm{rd}(G) \geq \lambda^+(G) = k$
and so $\mathrm{rd}(G) = k$.   ∎

Since the minimum size of a connected graph G of order n with $\mathrm{rd}(G) = k$ where
$1 \leq k \leq n - 1$ is now known by Theorem 8.2.8, this brings up the question of
determining the maximum size of a connected graph G of order n with $\mathrm{rd}(G) = k$.
Of course, we know this size is $n - 1$ when $k = 1$. Also, we know this size is $\binom{n}{2}$
when $k = n - 1$. For odd integers n, there is the following conjecture.

Conjecture 8.2.9 ([1]) *The maximum size of a connected graph G of odd order $n \geq$
5 with* $\mathrm{rd}(G) = k$, $1 \leq k \leq n - 1$, *is* $\frac{(k+1)(n-1)}{2}$.

Since $\frac{(k+1)(n-1)}{2} = n - 1$ when $k = 1$ and $\frac{(k+1)(n-1)}{2} = \binom{n}{2}$ when $k = n - 1$, this
conjecture is true for these two values of k. Also, $\frac{(k+1)(n-1)}{2} = \frac{3n-3}{2}$ when $k = 2$.
This is the size of the so-called *friendship graph* $\left(\frac{k-1}{2}\right) K_2 \vee K_1$ of order n (every two
vertices have a unique common neighbor). Since each block of a friendship graph is a
triangle, it follows by Theorem 8.2.4 that each such graph has rainbow disconnection
number 2.

For given integers k and n with $1 \leq k \leq n - 1$ where $n \geq 5$ is odd, let H_k be
a $(k - 1)$-regular graph of order $n - 1$. Since $n - 1$ is even, such graphs H_k exist.
Now, let $G_k = H_k \vee K_1$ be the join of H_k and K_1. Thus, G_k is a connected graph of

order n having one vertex of degree $n - 1$ and $n - 1$ vertices of degree k. The size m of G_k satisfies the equation

$$2m = (n - 1) + (n - 1)k = (k + 1)(n - 1)$$

and so $m = \frac{(k+1)(n-1)}{2}$. The graph H_k can be selected so that it is 1-factorable and so $\chi'(H_k) = k - 1$. If a proper $(k - 1)$-edge-coloring of H_k is given using the colors $1, 2, \ldots k - 1$, and every edge incident with the vertex of G_k of degree $n - 1$ is assigned the color k, then the edges incident with each vertex of degree k are properly colored with k colors. For any two vertices u and v of G_k, at least one of u and v has degree k in G_k, say $\deg_{G_k} u = k$. Then the set of edges incident with u is a $u - v$ rainbow cut in H. Since this is a rainbow disconnection k-coloring of G, it follows that $\text{rd}(G_k) \leq k$. It is reasonable to conjecture that $\text{rd}(G_k) = k$.

We would still be left with the question of whether every graph H of order n and size $\frac{(k+1)(n-1)}{2} + 1$ must have $\text{rd}(H) > k$. Certainly, every such graph H must contain at least two vertices whose degrees exceed k.

We conclude with a problem that combines two concepts in this chapter that have been investigated by Stephen Hedetniemi. A graph G is *vertex-colored* if every vertex of G is assigned a color from some prescribed set of colors. As with rainbow sets of edges, a set U of vertices in an vertex-colored graph is a *rainbow set* if no two vertices of U have the same color. A rainbow set U in a nontrivial connected vertex-colored graph G is a *rainbow cut* of G if U is both a rainbow set and a vertex-cut. For two nonadjacent vertices u and v of G, a rainbow cut U is a $u - v$ *rainbow cut* if u and v belong to different components of $G - U$. A vertex coloring of G is a *rainbow vertex disconnection coloring* if for every two nonadjacent vertices u and v of G, there exists a $u - v$ rainbow cut in G. The *rainbow vertex disconnection number* of a nontrivial connected graph G is the minimum number of colors required of a rainbow vertex disconnection coloring of G.

Problem 8.2.10 Investigate the rainbow vertex disconnection numbers of connected noncomplete graphs.

References

1. G. Chartrand, S. Devereaux, T.W. Haynes, S.T. Hedetniemi, P. Zhang, Rainbow disconnection in graphs. Discuss. Math. Graph Theory **38**, 1007–1021 (2018)
2. G. Chartrand, T.W. Haynes, S.T. Hedetniemi, P. Zhang, From connectivity to coloring. J. Combin. Math. Combin. Comput. **102**, 203–219 (2017)
3. G. Chartrand, G.L. Johns, K.A. McKeon, P. Zhang, Rainbow connection in graphs. Math. Bohem. **133**, 85–98 (2008)
4. P. Elias, A. Feinstein, C.E. Shannon, A note on the maximum flow through a network. IRE Trans. Inf. Theory IT **2**, 117–119 (1956)
5. L.R. Ford Jr., D.R. Fulkerson, Maximal flow through a network. Canad. J. Math. **8**, 399–404 (1956)
6. H. Hadwiger, Über eine Klassifikation der Streckenkomplexe. Vierteljschr. Naturforsch. Ges. Zürich **88**, 133–142 (1943)
7. X.L. Li, Y.F. Sun, *Rainbow Connections of Graphs* (Springer, Boston, 2012)

Printed in the United States
By Bookmasters